Design to Test

Design to Test

A Definitive Guide
for Electronic Design,
Manufacture, and Service

Second Edition

Jon Turino

VAN NOSTRAND REINHOLD
New York

Copyright © 1990 by Jon Turino

Library of Congress Catalog Number 89-25102

ISBN 0-442-00170-3

Printed in the United States of America

Van Nostrand Reinhold
115 Fifth Avenue
New York, New York 10003

Van Nostrand Reinhold International Company Limited
11 New Fetter Lane
London EC4P 4EE, England

Van Nostrand Reinhold
480 La Trobe Street
Melbourne, Victoria 3000, Australia

Nelson Canada
1120 Birchmount Road
Scarborough, Ontario M1K 5G4, Canada

16   15   14   13   12   11   10   9   8   7   6   5   4   3   2   1

Library of Congress Cataloging-in-Publication Data

Turino, Jon L.
    Design to test : a definitive guide for electronic design, manufacturing, and service / Jon Turino. -- 2nd ed.
        p.   cm.
    Includes bibliographical references.
    ISBN 0-442-00170-3 : $49.95
    1. Integrated circuits--Testing.   I. Title.
TK7874.T83   1990
621.381'5--dc20   89-25102
                                                                            CIP

This book is dedicated to Tina Nuss,
without whose love, help,
and support it would not exist.

Contents

List of Figures

Preface

This book is the second edition of *Design to Test*. The first edition, written by myself and H. Frank Binnendyk and first published in 1982, has undergone several printings and become a standard in many companies, even in some countries. Both Frank and I are very proud of the success that our customers have had in utilizing the information, all of it still applicable to today's electronic designs. But six years is a long time in any technology field. I therefore felt it was time to write a new edition.

This new edition, while retaining the basic testability principles first documented six years ago, contains the latest material on state-of-the-art testability techniques for electronic devices, boards, and systems and has been completely rewritten and updated. Chapter 15 from the first edition has been converted to an appendix. Chapter 6 has been expanded to cover the latest technology devices. Chapter 1 has been revised, and several examples throughout the book have been revised and updated. But sometimes the more things change, the more they stay the same. All of the guidelines and information presented in this book deal with the three basic testability principles—partitioning, control, and visibility. They have not changed in years. But many people have gotten smarter about how to implement those three basic testability principles, and it is the aim of this text to enlighten the reader regarding those new (and old) testability implementation techniques.

Much of the material contained in this book has been collected over the past 20 years from countless people who have participated with me in conferences and exhibitions around the world. A special thanks to those of you who have participated in my seminars. You have taught me a lot and forced me to stay current while giving me the opportunity to meet you in the United States, France, England, West Germany, Denmark, Sweden, Israel, Holland, Japan, Taiwan, and India. This book would not have been possible without your past and continued support.

Experience since publishing the first edition of *Design to Test* has proven the critical need for incorporating testability features in electronic designs, particularly as they become increasingly more complex. Experience has also proven the savings, in both money and time, that have been and can be achieved through proper design for testability. The examples sprinkled throughout this text are all real. The potential for savings in future designs is even larger.

The purpose of this book is threefold: to make the reader aware of the impact of testability, to define testability, and to give clear and simple guidelines so that each designer can implement *real testability* into his or her designs. The information contained here establishes guidelines which represent the collected experiences from many programs and the personal experiences of many managers, engineers, technicians, and manufacturing personnel throughout the world.

This book is *not* intended to provide a hard-and-fast set of rules to be implemented on every program or product design. Each program or situation is usually different. The intent is to provide each reader with ideas and approaches so that the end result of the design effort is a product that can be manufactured, tested, and maintained at minimum cost.

As I have pointed out in Chapter 1, design for testability is no longer just a "nice to have" attribute—it is a competitive necessity to increase the competitiveness of your new designs. Others are implementing the advice contained in this book. If you do not, you will not be as competitive as you could be, which would not bode well for your future. Strong words, perhaps, but true nonetheless.

Finally, I thank you, my readers, for acquiring what I hope will become your new "bible" on testability.

Acknowledgments

This book holds the accumulated knowledge of a great many people with whom I have become acquainted, and many times friends with, over the past 20 years. All deserve thanks and acknowledgment, and some deserve special mention. I'd like to thank Robert Linderman, who in 1979 trusted me enough to award my fledgling new company its first consulting contract (from DuPont). H. Frank Binnendyk, who helped write the book from which this second edition was created, also deserves special mention, as do Jim Shrmack and Richard Muir (of Raytheon), who gave us the original contract to build their internal manual on design for testability.

John Shearsmith in England helped to organize many seminars for us in the 1980s in England, France, and Germany that contributed to our knowledge of testability and our feeling for its internationality. Phil Koehler of UTI Corporation (and Bruce, Gordon, Nick, Case, Charlotte and Cindy) believed enough in us to help get the development of our testable functional circuits funded. And thanks to Mike Stora for helping co-found the IEEE Testability Bus Standardization Committee with me in 1986.

Acknowledgment is also deserved by Paul Bardell, Rod Tulloss, Gordon Robinson, and Colin Maunder for teaching us a great many lessons about boundary scan and the IEEE standardization process. Those lessons will not be forgotten. Our thanks go to Ray Chapman, Fred Harrison, Cliff Fowkes, and Tom Williams for carrying on the important testability bus standardization activities that augment scan techniques and make inherent testability and built-in test easier to implement.

There are a great many others who deserve acknowledgment

and thanks for their help and support over the past years and into the future. Barry, Trish, Peter, Rita, Graham, Maurice, ... Your names are far too many to list here, but I will remember you always. If I haven't told you recently how special you are, I promise to do so very soon.

1

Introduction

The increasing complexity of new products and the proliferation of new electronic device fabrication and packaging technologies used to implement each succeeding new design have made testability a necessary product performance attribute. For without testability, the most technically elegant product, from a "functions per square of space" standpoint, is absolutely useless.

It does absolutely no good to shave a day from design—ignoring testability—if that lack of testability adds weeks or even months to time to market. It must be possible to "design verify" and debug a new product design in the shortest possible time. It must also be possible to bring that product reliably to market in a competitive manner. That means that test programs must be generated to detect all of the possible faults that can occur in the product, both during product manufacturing and during the product's service life. It must also be possible to generate those test programs in a timely and efficient manner.

"Time to market" is a concept not always understood in the same way by different people. Some consider time to market to be raw schematic capture and design verification (e.g., good circuit simulation or prototype debug) time. Others realize that true time to market is the time it takes from the beginning of design until the product can be successfully delivered into the customer's hands at a competitive price. Figure 1-1 illustrates the effect proper testability design can have on product time to market.

Design for testability is no longer just a "nice to have" feature in a product. Nor is it strictly an engineering discipline. It is an element in a strategy of maintaining competitiveness in world markets, especially in the future. We can no longer continue to add cost to products through higher than necessary test programming times, test times, troubleshoot-

Product Designed <u>Without</u> Proper Testability Features

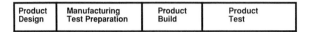

Product Design	Manufacturing Test Preparation	Product Build	Product Test

Product Designed <u>With</u> Testability

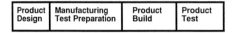

Product Design	Manufacturing Test Preparation	Product Build	Product Test

---

FIGURE 1-1. Time to market. Both design verification and test program generation time can be cut 5 to 15 percent when designs are made more testable from the beginning.

ing times, and high capital equipment costs. Technology exists to build testing resources into new product designs, thus drastically reducing the percentage of product cost made up by test and testing-related purchases and activities. Unless it is a built-in test that benefits the customer as well as the producer, test adds no value to a product; it just adds cost. Examine the simple real example in Table 1-1. The table illustrates the shift in parts, labor, and test costs that has occurred over the last five years for a typical (e.g., 100 medium scale integration (MSI) integrated circuit (IC) equivalent) mostly digital board.

While parts costs have decreased due to improvements in product yields and increasing levels of integration, and while assembly labor costs have decreased due to the use of fewer components per board design and increased levels of assembly automation, test costs have not decreased at all. Thus, while total product cost has decreased by 40% (from $500 to $300), test cost has risen as a percentage of product cost from 20 percent to 33 percent, even though its absolute cost may not have changed.

In many cases, however, the situation is much worse. As complexity increases and new packaging technologies (e.g., surface mount and fine pitch) continue to shrink parts count and labor content in new products, the actual dollar cost for testing has increased. This results in a test cost contribution to product cost of 35 percent to 55 percent (or more!), depending upon product size, technology, and complexity. That is a lot of added cost.

How does one control the rapidly rising cost contribution of test and testing-related activities? Simply by making the unit under test (UUT) testable! Take as an example the revolution in computer technology. Computers continue to get smaller, more powerful, and less expensive. Testers, on the other hand, tend to get larger, incrementally less

TABLE 1-1.   Test Cost as a Percentage of Product Cost

Cost	1983		1988	
	Actual ($)	Percent	Actual ($)	Percent
Parts	300	60	150	50
Labor	100	20	50	17
Test	100	20	100	33
Total	500	100	300	100

powerful, and far more expensive. Why is there such a dichotomy? Simply because the design of new products does not allow the testers to get smaller, more powerful, and less expensive! Most people are still using brute force (i.e., complex, expensive test equipment) to overcome unit under test testability deficiencies. The key, however, lies in the *prevention* of testability problems. That is where the leverage is.

Testability is a concerted corporate effort to reduce product costs and improve productivity and quality throughout the total business cycle—from product concept through design, manufacture, and usage in the field. Testability is not an attempt to restrict engineering innovations or to criticize the ability of the design to perform its function. We just want the ability to test the best (from a functional point of view) products in the least amount of time at the lowest cost. That means that future products must be testable.

HOW (AND WHY) CIRCUITS ARE TESTED

To understand the importance of testability, we must understand how and why circuits are tested. Circuits are tested by applying digital or analog stimulus signals to circuit input pins and verifying the response of the unit under test to those signals by evaluating the response signals of the unit under test. In the digital realm, the input stimulus signals, usually called *test vectors*, and the resulting response signals are patterns of logic 1's and 0's. In the analog world, the input stimulus signals, and thus the resulting responses to be analyzed, can be quite complex variations of frequency, voltage, resistance, or other parameters.

Circuits are tested in order to detect all of the possible faults in a unit under test that could prevent proper circuit operation. Faults can occur in the components used in an assembly or as a result of the assembly process. Faults within components are usually referred to as *functional faults*. Faults that occur on the assembly are usually referred to as *manufacturing defects* or structural faults. Faults that occur due to functional interaction problems between good components on a good assembly are *design defects*. Faults that occur after an assembly or system has been placed in service are usually functional faults.

In the digital realm, faults are detected by applying stimulus vectors (i.e., sets of logic 1's and 0's) to circuit input pins in an attempt to cause every node (i.e., circuit interconnection) in the circuit to be at the logic 0 state at least once and to be at the logic 1 state at least once. This is called *fault activation*. But it is not enough.

Faults must also be propagated to circuit output pins to ensure that each node actually assumed the state that it was ordered to assume by the stimulus vectors. Faults are propagated by providing a path from the circuit node being activated through the other circuits in the design to the circuit's physical output pins (or test points). *Fault detection thus requires activating faults and propagating faults.*

The analog world uses a similar procedure but different input signals and output response analysis methods. Analog stimulus signals are applied to circuit input pins in order to exercise each circuit node to its full range of parameters. This is analogous to the digital logic 0 (low) and logic 1 (high) states. Analog responses are measured parametrically (i.e., volts, amps, time, frequency, and waveform characteristics) in order to determine if the node under test actually did what it was supposed to do.

The concept of fault activation and fault propagation is illustrated in Figures 1-2 and 1-3 with a digital circuit example. In Figure 1-2, the objective is to answer the question "Is U5's output stuck at the logic 1 state?" To answer that question, we must place test signals on the inputs to cause U5's output to go to the logic 0 state (if it is working).

In Figure 1-2, U5's output will go to the logic 0 state when both of its inputs are sent to the logic 1 state. To provide a logic 1 state at the output of U1, its inputs must both be at the logic 1 state. To provide a logic 1 state at the output of U2, both of its inputs must be in the logic 0 state. Thus the two logic 1's and the two logic 0's at the upper left corner of the figure make up the portion of a test stimulus vector that will activate the fault "U5 stuck at 1." In other words, if U5's output is not "stuck" in the logic 1 state (i.e., it is not faulty), it will assume the logic 0 state with the test stimulus vector so far applied.

The testing job is not yet complete. Although we have created a set of conditions that will cause U5's output to change to the logic 0 state in response to a test stimulus vector, we have no way of observing whether U5's output *actually went* to the logic 0 state. There is no known path through U7 to allow us to see the results of the input stimulus to U5.

Logic 1's and 0's must be applied to the inputs of U3 and U4 to

Q: Is U5 Output "Stuck" at Logic 1?

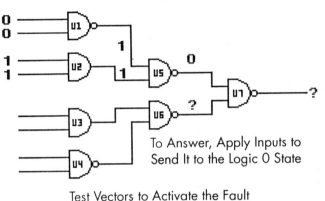

To Answer, Apply Inputs to
Send It to the Logic 0 State

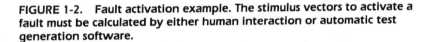

Test Vectors to Activate the Fault

FIGURE 1-2. Fault activation example. The stimulus vectors to activate a fault must be calculated by either human interaction or automatic test generation software.

provide the correct states to U6 to send its output to the logic 1 state as well. With U6's output at logic 1, U7 will transfer the results of activity on U5's output to the circuit output pin. The test vector is now complete, and we can detect a "stuck-at-1" fault at the output of U5.

This test vector detects quite a few faults in addition to U5 stuck at

Q: Is U5 Output "Stuck" at Logic 1?

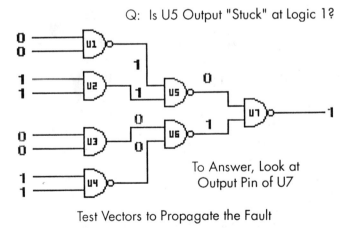

To Answer, Look at
Output Pin of U7

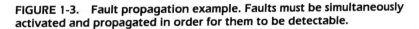

Test Vectors to Propagate the Fault

FIGURE 1-3. Fault propagation example. Faults must be simultaneously activated and propagated in order for them to be detectable.

1. It detects input faults stuck at 0 for U1 and U3, input faults stuck at 1 for U2 and U4, output faults stuck-at-0 for U1-U4, input stuck at 0 faults for U5 and U6, output stuck at 0 for U6, the top input of U7 stuck at 1, the bottom input of U7 stuck at 0, and the output of U7 stuck at 1. If U7's output does not respond correctly to the input test vector, where is the fault? That question brings up the concept of "ambiguity groups"—if there is a fault, which component is actually causing the fault?

Diagnostic accuracy may not be important when the part being replaced costs only a few dollars, but it is critically important with expensive parts such as application specific ICs (ASICs), microprocessors, and other complex very large scale integration (VLSI) devices. As rework costs continue to increase, diagnostic accuracy is especially important for surface-mounted components.

This example is admittedly trivial, but it was presented to illustrate certain concepts. Consider what happens with a much more complex circuit, such as the one in Figure 1-4. This circuit is composed of 20 ICs, some VLSI ICs, and some ASICs, along with regular "glue logic." It may contain feedback loops, it may or may not be initializable, there may be long data paths or counter chains between the physical inputs and outputs, and there are many levels of logic to be stimulated in order to both activate and propagate all of the possible faults that could prevent proper circuit operation. This circuit is extremely difficult to test unless testability features are added.

Consider only the task of trying to find one of the thousands of

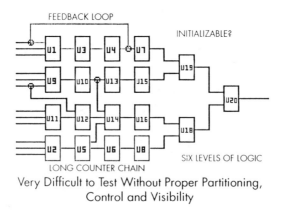

Very Difficult to Test Without Proper Partitioning, Control and Visibility

FIGURE 1-4.   A more complex example. While it is relatively straightforward to activate and propagate faults in simple combinatorial circuits, the job can become very difficult with complex circuits that include sequential logic.

possible faults in this circuit—that of IC U3's output stuck at 1 (or 0). It may or may not be too difficult to activate the fault by stimulating U1 according to its truth table or functional behavior, depending on whether what is done to U3 affects U4 in such a way as to feed a signal back to U1 that conflicts with the desired applied stimulus vectors. But even if the feedback loop does not present a problem, propagating that fault to the circuit output (at the output of IC U20) requires that all of the other circuitry also be stimulated—no longer a trivial task.

KEY TESTABILITY TECHNIQUES

In order to make complex designs testable, three key testability principles, implementable in a great many ways, must be included in the circuit design. They are:

- Partitioning
- Controllability
- Visibility

We partition circuits by breaking them into reasonably small functional blocks, or clusters. This makes them easier to understand, easier to write tests for, and easier to test and troubleshoot. We provide circuit control by including *reasonably direct* paths from the test resource (either automatic test equipment or built-in test circuitry) to critical internal nodes required for initialization of the circuitry under test, partitioning of that circuitry, and control for fault activation. We provide circuit visibility by bringing internal nodes to the testing interface, again in a reasonably direct manner. This principle reduces logic and fault simulation times and costs along with test generation, testing, and troubleshooting times.

These principles are easy to implement with minimal effect on circuit configuration, performance, and reliability. The circuit of Figure 1-5 shows three gates added to Figure 1-4 to partition it into smaller functional blocks. This allows certain blocks of the circuitry to be stimulated while leaving other circuit elements in the inactive state (i.e., not needing to be stimulated to activate faults in the subsequent circuitry).

The added gates may, in some cases, add an extra delay between circuit elements. In other cases, no delay is added. The circuit is simply modified to include an extra input (called an extra fan-in point) to directly control a circuit function for testing purposes. This extra input drastically cuts the time needed to generate the stimulus vectors necessary to activate faults.

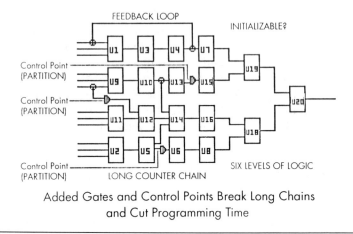

Added Gates and Control Points Break Long Chains
and Cut Programming Time

FIGURE 1-5.   Partitioning circuits into smaller blocks. Extra gates and inputs allow the circuit to be segmented into smaller "clusters" that can be dealt with individually.

Large sequential circuits are also not always initializable (i.e., able to be set immediately to known states for testing) and often contain feedback loops and long counter chains. A few strategically selected control points allow for direct (immediate) initialization of memory elements and other sequential circuits. They can also be used, usually with an extra gate or two, to break feedback loops. Examples of these types of test control points are shown in Figure 1-6.

A reset line has been added to provide immediate initialization, and an extra input has been provided to allow the feedback loop to be disabled, eliminating any conflicts between the desired applied stimulus patterns at the inputs to IC U1 and the results of those patterns on the outputs of U4.

Lack of initialization and the presence of feedback loops contribute significantly to long test generation, logic simulation, and fault detection times (and resulting higher costs).

To further reduce times and costs, visibility (or observability) points are added so that extra stimulus patterns are not needed in order to propagate faults to circuit output pins via other complex circuits. With visibility points added, faults are propagated immediately to the tester or built-in test interface. Figure 1-7 shows the addition of four visibility points to the already partitioned and controllable complex example.

Consider now, with the addition of four simple gates, five input

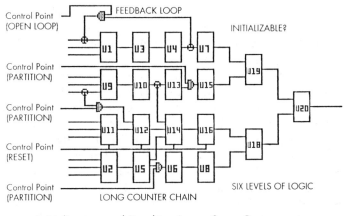

Initialization and Breaking Loops Saves Programming
and Troubleshooting Costs

FIGURE 1-6.   Providing control point access. Inputs and gates for initializing circuitry and for breaking feedback loops prevent the propagation of unknown logic states and conflicts between stimulus vectors and other circuit operations.

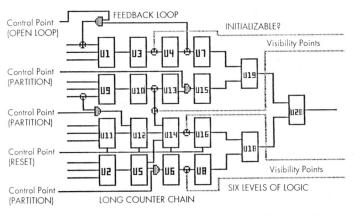

Visibility Reduces Test Programming, Simulation and Troubleshooting Times

FIGURE 1-7.   Adding visibility points. Direct paths from internal circuit nodes to card edge connector pins further segments the circuit into small blocks. Compare finding a fault on the output of U3 with the one shown in Figure 1-4.

control points and four output visibility points, the problem of activating and propagating the fault "U3 stuck at 1(0)." It is an infinitely simpler problem to solve now that the circuit has been partitioned and made controllable and visible.

IC U3 can be directly stimulated through U1 without worrying about the effect of U3's outputs on U4. The actual output of U3 can be directly and immediately observed by the tester or built-in test circuitry without stimulating all of the surrounding, or intervening, circuitry. In short, the use of partitioning, control, and visibility significantly lowers all times and costs associated with both design verification and test. And, as shown, their implementation requires very little in terms of extra circuitry or extra circuit input/output connections.

Testability techniques are designed to reduce, or at least control, ever-escalating test costs. Product design engineers must design circuits that can be tested in an efficient, economical, and orderly manner. Incorporating features that facilitate testing and fault isolation and that help reduce maintenance costs over the life cycle of the product is the responsibility of the product designer and should be part of the specification for each new product.

If you are the customer, require testable designs from your suppliers. If you are the producer, look at testability not as a burden to your product designs but as a feature that will make your product more attractive to the potential user. If you are the designer, make sure that your product, however technically elegant, can be produced, tested, and serviced at a competitive cost. Otherwise, it won't be as successful as it could be.

TESTABILITY DEFINITIONS

Testability should be thought of, from a management standpoint, as a concerted corporate effort to provide maximum efficiency and economy throughout the *total business cycle*, from product concept through design, manufacture, and service. Testability is not an attempt to restrict design innovations or to criticize the efficacy of a design to perform its intended function.

In quantitative terms, testability is defined as a measure of the ease with which comprehensive test programs can be written and executed, as well as the ease with which faults can be isolated on defective components, subassemblies, and systems. The higher the testability of a product, the lower is its overall cost and the greater is its competitiveness.

In technical terms, testability is *partitioning*, *control*, and *visibility*.

Why isn't testability implemented as thoroughly as it should be?

Testability techniques and benefits have been documented for years. Perhaps the answers lie in the (previously possibly true but no longer valid) misconceptions that implementing testability will have a negative impact on design cost and design schedules.

Especially in these days of ASICs and designers being required to perform design verification and logic (and sometimes fault) simulation, implementing testability techniques will actually reduce the design cost and shorten the design schedule.

Perhaps the answer lies in worries about the impact of testability on product reliability. Reliability, either calculated or actual (or both), is a critical parameter in most system design specifications and procurement. Actual reliability, and not calculated reliability, is a critical element in the deployment and use of electronic systems. Maintainability, again both calculated and actual (but actual maintainability is more important), is also a critical element in most specifications, procurements, and use.

To make something maintainable, one must make it testable. To make something testable, one may have to increase the numbers or complexity of the components that go into it. If one increases the number or complexity of the components that go into a system in order to make it testable and maintainable, the *calculated* reliability figures may go down; however, the actual reliability may not decrease.

Is absolute (either calculated or real) reliability what most users are looking for? Or is *availability* really the critical factor? Availability (i.e., real availability, which may or may not be related to calculated availability) is the figure of merit that ties together reliability, testability, and maintainability. Consider the following example: A system has a mean time between failures (MTBF) of 100 hours and a mean time to repair (MTTR) of 10 hours. It is thus "available" 90% of the time.

If we make the system more testable and easier to repair by adding a 1% component count/component complexity increased burden to it, we might make it less reliable with an MTBF of 99.9 hours. The testability and maintainability improvements, however, have reduced the MTTR from 10 hours to 1 hour. The system is thus "available" 99+% of the time (compared with the previous 90%). Which system figures would you rather have? With the exception of aircraft engines, almost every other system will benefit from improved testability and maintainability, even at the expense of calculated reliability figures. And, in practice, testability improvements to achieve the kind of reduced MTTRs just mentioned have virtually no impact on actual reliability, although they may have a small impact on calculated reliability.

It is true that every added component and every added connection in an electronic system design degrades reliability. In earlier days, when achieving testability meant adding many discrete components and many extra dedicated connections to a design, the reliability impact was sometimes significant. In this day and age of large scale and very large scale integration (LSI/VLSI) ICs and full custom and application specific ICs, however, testability can be easily implemented without undue degradation of reliability. The advances in technology have given designers the ability to implement true functional testability with virtually no impact on reliability. Only the old belief that testability and reliability are at odds remains.

Perhaps it is the worry of the design engineer that implementing testability will have an adverse impact on circuit performance. Certainly some testability techniques do adversely affect circuit performance (and other circuit parameters, such as weight, size, and power consumption). But there are equally as many, if not more, state-of-the-art testability techniques that have absolutely no impact on circuit performance and absolutely minimal effect on other circuit parameters (e.g., weight, size, and power consumption).

The trick is in being smart enough to implement the right testability technique in the right design situation. In an ASIC, use one of the many scan techniques. At board level, choose testable parts rather than typically untestable single-function parts when implementing "glue logic." At the system level, architect the system to take advantage of built-in tests and make sure that all subsets of the system are inherently testable (i.e., partitioned, controllable, and visible).

Perhaps the real reason, however, that testability has not been implemented as thoroughly as it should have been is more of an organizational problem than a technical problem. The benefits of incorporating testability in a design, though requiring a little extra effort by the design engineer (or, more accurately, more attention, since many testability features can be automatically inserted into new designs by computer-aided engineering workstation resident software packages), appear to accrue to the production, test, quality, and service organizations.

But aren't production, test, quality, and service all part of the same business that the designer contributes to as well? Of course they are! It is time to break down the "walls" between the various groups in a business. It is time to make sure that cooperation and commitment to the success of the business are more important than organizational rivalries and power struggles. It is time to stop focusing on part costs alone and start focusing on *overall product costs*. It is time to make things testable enough so that we can adequately verify their quality. It is time to make sure that top management communicates the message to all of the elements in the organization that products must be manufacturable, testable, and serviceable if they are to be competitive in future world electronics markets. It is time for real functional testability. Those who ignore the testability precepts outlined in this text will find themselves at a considerable competitive disadvantage in the years to come.

WHY IS TESTABILITY IMPORTANT?

There are many answers to the question, "why is testability important?"

- To improve design quality.
- To improve product quality, availability, and acceptance.
- To decrease time to market.
- To decrease production test, field service, and fault simulation costs.
- To decrease capital equipment investments for automatic test equipment.
- To eliminate test bottlenecks.
- To reduce design verification time.
- To reduce organizational strife between design, test, and management.
- To increase market share, profits, and cash.
- To survive as a viable electronics manufacturer.

Paying attention to testability during the design stage of a product's life cycle and enforcing the implementation of the proper testability implementation techniques is a high-leverage activity. Even with shorter and shorter product life cycles, the extra hours or days spent making a design testable and the extra dollar or two in parts costs per product will make a big difference in the overall costs of staying in business.

The savings from proper testability implementation in a low-volume, wide-variety business typically come in the nonrecurring engineering areas—expenditures for test equipment, test programs, and test fixtures (along with the relevant documentation). Table 1-2 shows one example where a relatively large testability investment (5% added to design cost and 4% added to product parts costs) still results in a product cost reduction of over 15%.

In a low-variety, high-volume operation, the major savings are realized in the recurring cost areas—ongoing testing and troubleshooting costs (see Table 1-3). For most people, savings occur in both nonrecurring and recurring cost centers. In any case, the savings are large and significant.

TABLE 1-2.   Low-Volume, Wide-Variety
Economic Example

Major Cost Category	Without Testability ($)	With Testability ($)
Design (per qty.)	93.54	98.90
Parts (each)	273.16	285.32
Programming (PQ)	147.75	85.54
Fixturing (PQ)	24.68	7.63
Testing (each)	4.83	3.67
Diagnostic (avg. ea.)	19.14	9.18
Rework (avg. ea.)	15.86	12.69
Test Equipment (PQ)	18.37	11.74
Total Product Cost per Item	597.33	514.67
Savings/Year = $82.66 per board × 5,000 boards = $413,300.00		

Notes: Low-volume, wide-variety example; boards contain
approximately 150 ICs each, 25 new designs per year, 200 of
each type built per year, ATE amortized over 5,000 boards,
other items over 200 boards.

TABLE 1-3.   High-Volume, Low-Variety
Economic Examples

Major Cost Category	Without Testability ($)	With Testability ($)
Diagnostic (per qty.)	3.97	4.12
Parts (each)	148.56	157.61
Programming (PQ)	3.29	1.68
Fixturing (PQ)	0.46	0.12
Testing (each)	2.93	2.64
Diagnostic (avg. ea.)	16.65	7.63
Rework (avg. ea.)	14.34	12.15
Test Equipment (PQ)	9.80	5.30
Total Production Cost per Item	200.00	191.25
Savings/Year = $8.75 per board × 50,000 boards = $437,500		

Notes: High-volume, low-variety example; boards contain
approximately 150 ICs each, 5 new designs per year, 10,000 of
each type built per year, ATE amortized over 50,000 boards,
other items over 10,000 boards.

TESTABILITY AWARENESS

Testability is *not* a technological innovation in itself, although many technological innovations in its implementation have taken place. Testability awareness is a way of thinking wherein the *designer possesses an awareness of the importance of testing.* Testability is then designed in at the proper time—from the start. The proper time is when the component, printed circuit board (PCB), or system is initially being specified and/or designed.

Testability should be part of the design cycle because testing is a function that must be performed during the business cycle. Testability should be thought of as part of the functional specifications that the design must meet. It is crucial that, before starting any design, the designer asks: "How will I test it, how will production test it, and how will it be tested in the field?" "Which testability implementation technique will I use at each level to ensure that my design is successfully testable?"

TESTABILITY COMMITMENT

This book is designed to assist anyone involved with or responsible for testing, but it is especially for the product designer because the designer holds the key to the success of any program. One of your major responsibilities, if you are a design-oriented person, is to design products that are efficiently and economically testable. This book will enable you to accomplish that goal in three ways:

1. Utilization of the data in this book will provide you with the knowledge you need to enhance the testability of your design.
2. Noting down and improving the data in this book will enable you to continue to improve the usefulness of this book as a testability tool.
3. Providing feedback within your organization about the implementation of the guidelines in this book will make everyone's job easier, more productive, and more enjoyable.

TESTABILITY BENEFITS

The benefits of designing for testability can be realized both by the individual and by the entire organization. Here are some:

- It reduces the time required to transfer a design from design engineering to manufacturing engineering.
- It reduces post-release involvement of the design engineer to get a design smoothly incorporated into the production line.
- It reduces manufacturing cost and increases profits.
- It improves the working relationship between design engineering and manufacturing (test) engineering personnel.
- It produces products with lower initial and life cycle costs. This will serve to increase product sales and market share.
- It decreases test times and reduces production delays.
- It increases field service productivity by allowing for more efficient diagnosis and repair.

In summary, when finished with this book you will realize that real functional testability can be implemented with a very small impact on product design. The improvements you make to your product designs will allow those products to be tested with far fewer test patterns and in much shorter times. Expensive capital test equipment will last longer and provide higher fault coverage (i.e., quality) with reduced test generation times and unambiguous fault isolation.

Design for testability techniques must be implemented in new product designs if tomorrow's new products are to be testable on time and at a competitive cost. The pressure is even more clearly evident when one examines testability trends for the future.

TESTABILITY TRENDS FOR THE FUTURE

New technical trends can make testing even tougher. Just as test and design engineers begin to get a handle on testability, along come ever more complex devices, many of them without adequate built-in testability. When these new, more complex devices are assembled into or onto printed circuit boards, the result can be an almost completely untestable design.

On the plus side, there is more networking going on between computer-aided engineering (CAE), computer-aided design (CAD), and computer-aided test (CAT) equipment. That networking means that a lot of data that previously had to be created twice can now be directly (electronically) transferred.

On the negative side, new packaging techniques, particularly surface mount technology and "chip-on-board" and fine pitch technology are making it possible to pack even more complexity into the same area of board space. Increased device complexity is permeating all types of semiconductors—standard, semicustom, and full custom.

Commercial parts, especially special-purpose devices like communications chips and peripheral controllers, tend to gain complexity each year. And while the device manufacturer may be able to test them, the user has a particular problem when they are mounted on boards.

The increased use of user-programmable devices, gate arrays, full-custom ICs, and semicustom (or application specific) ICs means that many new devices are being designed by many designers who are totally unaware of testability techniques. Some "silicon foundries," however, are making sure that their customer designers include some on-chip built-in test.

The same increase in complexity that is characteristic of semiconductor devices drives increased complexity at the PCB level. More complex devices are populating new board designs. Often, lines provided on the devices for self-test or built-in test are not accessible once the board has been designed.

The trend toward larger boards, in the mistaken assumption that a large board saves cost, is also hurting productivity. Although a large board does save on parts costs, it tends to cost a lot in the other business cost elements, most notably test.

Whole groups of SSI/MSI devices, once at least accessible with bed-of-nails fixtures (Figure 1-8), are now being encased in plastic or ceramic as gate arrays or ASICs. And new manufacturing techniques, such as surface mount technology, are letting designers quickly design boards so complex that they potentially cannot be manufactured and tested at all!

As mentioned earlier, however, there are some advances being made in the data base and factory networking areas. CAD/CAE/CAT tie-ins allow for the direct electronic transfer of much of the data re-

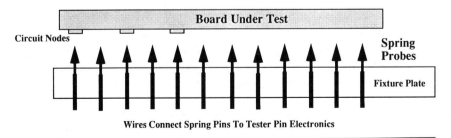

Wires Connect Spring Pins To Tester Pin Electronics

FIGURE 1-8.   Bed-of-nails fixture. A bed-of-nails fixture uses spring-loaded probes to connect to circuit nodes on the solder side of a PCB.

quired for effective and efficient testing to the test equipment and/or for the test engineer.

Most automatic test equipment (ATE) can now accept at least net lists from the CAD system. Newer systems allow for the transfer of actual circuit models and test patterns, many of which are now being generated by the designer on the CAE/CAD system as part of the design verification process.

Connecting the ATE into a factory network allows test and quality personnel to improve manufacturing yield and to feed information back to design on hard-to-manufacture, hard-to-test, or unreliable designs.

Perhaps the most dramatic impetus currently being given to testability is the push from new packaging techniques. Surface mount components, surface mount technology (SMT) boards, and chip-on-board techniques are all possible ways to squeeze more performance out of the same product size. Hampering the widespread use of these techniques, however, is the lack of information on how to test these new assemblies. As components get placed on both sides of the boards, and because the components themselves have leads spaced very close together (often well under even the 0.050-inch level that expensive bed-of-nails fixtures can access), the traditional in-circuit testing approach is experiencing problems. Since more automation is required for SMT manufacturing, the fault spectrum on boards is changing. Fewer manufacturing defects, the in-circuit tester's forte, will be encountered.

To summarize the result of all of these trends and advances is quite simple. As shown above, more requirements for built-in test and testability will be imposed on the circuit designers. If they are not, the new designs will not be testable.

Parts will have to be added or selected properly for SMT board designs to make test programming and fault isolation affordable again. The economics of adding a few parts—using a square inch or two of board space—for testability and built-in test will prove irrefutable in the face of reduced product costs.

Also, with the cost of testing, troubleshooting, and rework rising as board density increases, and as it becomes increasingly difficult to replace wrong, wrongly oriented, and defective components, more attention must be paid to manufacturing as a process rather than as a "batch" operation and to controlling that process so that only good products get built.

In summary, the trends for the future are:

- More attention to design for testability techniques and built-in test approaches.
- More attention to manufacturing process control.

DESIGN-TO-TEST OVERVIEW

Design-to-test information contained in this book encompasses the latest in system design guidelines; digital, LSI/VLSI, analog, mechanical, and software design guidelines; evaluation procedures; testability documentation; automatic test equipment techniques; and checklists. Each of these major sections is explained next.

System level guidelines (Chapter 2) contains ideas and suggestions such as built-in test equipment (BITE), partitioning, and human factors for making a product more testable. This section encompasses electrical, mechanical, and software technology.

General digital circuit guidelines (Chapter 3) contains ideas and suggestions such as increased visibility and control for all SSI and MSI equivalent (i.e., gate array, PAL, and custom IC) circuit designs that make digital designs more testable.

General analog circuit guidelines (Chapter 4) contains ideas and suggestions such as increased visibility, control, and partitioning for making analog designs more testable.

LSI/VLSI board-level guidelines (Chapter 5) contains ideas and suggestions such as increased visibility, control, and built-in diagnostics for making LSI/VLSI circuitry more testable.

Merchant devices on boards (Chapter 6) contains detailed guidelines regarding which input and output pins of commercially available microprocessors and support circuits should be made controllable and visible.

LSI/VLSI ASIC level techniques (Chapter 7) covers the concepts of structured design for testability and the techniques that are most often implemented in custom and semicustom LSI/VLSI integrated circuits.

Boundary scan (Chapter 8) explains the concepts and the details of this on-chip approach that is aimed at easing future board level manufacturing defects testing problems.

Built-in test techniques (Chapter 9) explores the objectives of built-in test techniques, looks at the available alternatives and illustrates several built-in test approaches.

Testability buses (Chapter 10) looks at the concept of formalized testability interfaces and covers both existing and proposed standards for testability bus implementations.

Mechanical testability guidelines (Chapter 11) contains ideas and suggestions such as accessibility, identification, partitioning, and human factors for making mechanical designs more testable.

Surface mount technology guidelines (Chapter 12) lists the major mechanical and electrical concerns and guidelines related to surface mount and fine-pitch technologies.

Software guidelines (Chapter 13) contains ideas and suggestions such as single-entry and single-exit subroutines, diagnostic routines, and partitioning for making software more testable.

Testability documentation (Chapter 14) contains guidelines for the generation of test procedures, calibration procedures, and other relevant documents.

Testability implementation (Chapter 15) defines methods by which designers and managers can ensure that proper consideration is given to testability in system, hardware, and software design.

Test techniques and strategies (Chapter 16) provides a list of the capabilities, uses, and limitations of most standard automatic test equipment in use today in industry. The capabilities of this equipment are constantly being improved to deal with new technologies and the needs of design engineering, manufacturing, and management personnel.

Testability checklists (Appendix A) are provided to help you double-check that a thorough implementation of good testability techniques has been achieved. The checklist section is a hybrid of industry standard and U.S. MIL-STD-2165 checklists.

Digital T-score rating system (Appendix B) outlines an actual procedure for calculating a testability "figure of merit" for digital circuit board designs.

Appendix C includes a list of references and credits for further study of any testability area.

2

System Guidelines

This chapter contains ideas, methods, and techniques for improving the testability of complete products, whether they are one-board products or multiple-board and multiple-subassembly products, at what is termed the system level. Testability does not just happen; it must be planned from the top down, right from the beginning of the system specification and architecture determination phases of product design.

Each program or product must be considered with respect to the restrictions placed on it in terms of technical and performance requirements, size, weight, power consumption, reliability, and so on. Then the required testability features must be identified and merged into the system architecture (or vice versa) and the various testability implementation circuitry segments allocated to the appropriate elements of the system (i.e., into the ICs, boards, and subsystems that make up the product).

To determine the approach for any given program or product, the following requirements are major factors to be considered in the decision-making process:

- Built-in test (BIT) and built-in test equipment (BITE)/self-test features, requirements, and benefits
- Factory test flows and production volumes
- Maintenance and repair concepts, methods, equipment, and personnel skill levels
- Spares philosophy (including repair of returned spares)
- Mean time between failure (MTBF), mean time to repair (MTTR), and customer equipment availability requirements
- Environmental requirements including size, packaging technologies, and so on

- Final product applications/use
- Test equipment concept/availability/capability
- Skill levels of test personnel
- Special customer considerations and requirements

SYSTEM ANALYSIS

Systems should be analyzed for testability just as they are for functionality. Just as trade-offs are made during system analysis and architecting, trade-offs in the testability and/or built-in test aspects of the system should be made. Consider system analysis from both the electrical and mechanical points of view. System analysis with regard to testability must encompass, as a minimum, the following:

- System logic and circuit simulation (including fault simulation)
- Customer specified built-in test requirements
- System test and fault isolation approaches
- Production test and service troubleshooting aids (e.g., PCB extender cards)
- Printed circuit board testing and repair
- Allocation of testability and/or built-in test elements to the various levels of integration (i.e., on-chip BIT/testability, on-board BIT/testability, and in-system BIT)

Such simple-minded items as cable removal methods, number, types, and difficulty in making adjustments, select-on-test options, and module and subassembly removal and replacement are often forgotten until late in the product design stage. The result is, more often than not, a design that is difficult to manufacture, test, and diagnose. For if basic strategic testability decisions are not made right up front, it is very difficult (and sometimes impossible) to retrofit testability and maintainability features back into a design.

A testability technical plan should be developed for each program or product at the start of the design phase in order to make sure that the basic strategic decisions regarding testability do not take second (or last!) place in the designer's list of priorities and requirements. That plan should take into consideration at least the following factors:

- Number and types of connectors to be used
- Common power and ground pins
- Types, locations, and methods of interconnecting with test

points (e.g., spare "test only" IC sockets on PCBs versus extra edge connector pins or a testability bus connector)

- Number, types, and technologies of PCBs to be included in the system (e.g., single-sided/double-sided, two-layer/multilayer, with/without solder mask, with/without embedded resistors)
- System testability requirements (BIT/BITE)
- Extender card requirements
- Mechanical packaging requirements
- Test equipment requirements
- Test strategy recommendations and economic analyses, test equipment recommendations, considerations, and selection
- Fault isolation criteria (system level as well as number of PCBs and number of components on a PCB that must be isolated to with and without external test equipment)
- Operational and testability software requirements
- Special environmental requirements

The remaining portions of this chapter discuss additional strategic decisions that should be made at the system level and offer guidance for insuring that system designs include inherent testability features that facilitate both testing with external automatic test equipment (ATE) and BIT resources.

SYSTEM LEVEL TESTABILITY GUIDELINES

This section deals with general electrical testability ideas that can be applied to any electrically related discipline such as the design of digital, analog, RF/IF, or electronic systems. It begins with a brief discussion of BIT/BITE concepts, which are covered in detail in later chapters, and concludes with guidelines for insuring adequate inherent testability in the elements of a system when a coherent built-in test strategy is not appropriate for a specific design.

Removable BITE Concept

In many systems, fully integrated built-in test may sometimes unfortunately become a secondary (or tertiary) consideration due to the actual (or perceived) conflict between BIT needs and the system specification requirements for power consumption, reliability, space, and so on. Whatever the reason, one approach is to leave some room in the assemblies to insert pluggable cards which can be used as built-in test equip-

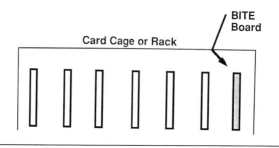

FIGURE 2-1. Removable BITE concept. If it is not possible to ship a built-in test board with each system, it may be possible to leave space for it when its use is required or desirable.

ment. These cards, the use of which is shown in Figure 2-1, can include a variety of functions, such as buffers, indicators, and connectors to external test equipment.

The BITE boards may be removed when they are not in use (i.e., during normal system operation). The space for them, however, cannot be removed during design for it will never be regained. Discipline must be employed to make sure that the space for removable BITE assemblies is not given up at the request of the designer (or even the customer) for a few more system features or functions.

BIT Concepts

Built-in test is becoming a mandatory feature in both commercial and military systems as we move into the 1990s. On the military side, U.S. MIL-STD-2165A, Testability Program for Systems and Equipment, is often invoked for new system designs, and it mandates testability. On the commercial side, the economics of manufacturing and a competitive world market posture require it. For without them, the costs of test program generation and troubleshooting will continue their very rapid rise from today's already lofty levels.

The following paragraphs introduce the concepts and economics of BIT in order to introduce the reader to BIT concepts before he or she delves into them in later chapters.

The basic premise behind built-in test is that "distributing" test circuitry on each board (or throughout each system) results in faster test program generation, quicker product confidence testing, and, perhaps most importantly, more efficient troubleshooting. The result is that total business costs are lowered through an early investment in testing during product design.

Most built-in test capabilities today focus primarily on the product confidence, or go/no-go, aspects of testing. Go/no-go BIT is very good for telling you if the system or product under test is functioning. It does not, however, usually provide any indication of where a fault may lie in the event that the system is nonfunctional or whether a fault in the unit under test prevents the BIT from providing any information.

Electronics manufacturers are only now beginning to realize the real benefits of including diagnostic (i.e., fault isolation) capabilities in the built-in test routines. The costs of test program generation and fault isolation are directly related to the level of control available over functional circuitry under test, and the amount of observability (or, more commonly, visibility) into the circuitry that exists. As control and visibility increase, test programming costs decrease, the tests can be carried out faster, and fault diagnostics are much faster and more accurate.

Figures 2-2 and 2-3 illustrate the concept of using some sort of testability bus (T-bus) to make boards and systems far more testable. In Figure 2-2, the connections made to on-board internal circuit nodes via the T-bus interface chip set circuits can be thought of as "windows" into the functional circuitry. Control and visibility access to the internal nodes are available not only through the (sometimes long and arduous) normal edge connector path but also through the testability bus connector. Adding testability interface circuits to a board, or replacing typically untestable single function glue logic circuits with dual function testable circuits that provide both functionality and testability, makes implementing BIT a far easier task.

In Figure 2-3, each board contains, in addition to its functional

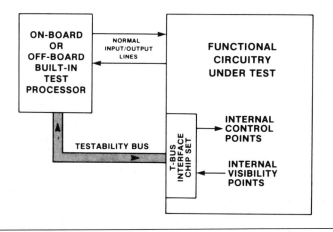

FIGURE 2-2.   Board level BIT example. Extra circuits added to a board can provide a testability bus interface that allows for more direct control and observation of internal circuit nodes to help with fault detection and fault isolation.

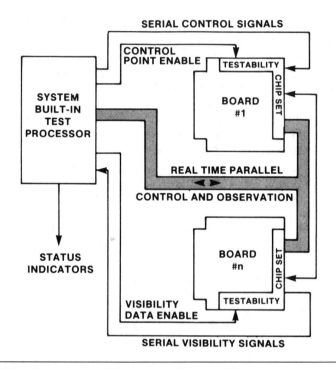

FIGURE 2-3. System level BIT example. The testability bus concept can be extended to the system level if all boards have a testability interface. The figure shows a dual-port bus with serial and real-time capabilities.

circuitry, a few control ICs and a few visibility ICs. The boards are connected to the system built-in test processor via the testability bus. The real-time control and observation signals are written to and read from each board's testability interface circuitry via an addressing scheme. The serial control and visibility signals connect to and from the testability bus at the first and last boards, respectively, and transfer data along their paths in a "daisy chain" fashion. Thus, there are two windows into the functional circuitry at all times. Should the system built-in test processor desire to initiate a self-test, it need only provide the appropriate data to the testability chip set and activate the ENABLE line.

Similarly, should the built-in test processor desire to monitor the activity of the internal nodes of the boards in the system, it need only decide whether to look at them serially or in real time and provide the appropriate data to accomplish the selected task. A 25-pin testability bus can access up to 4,096 internal board nodes on as many boards as the system contains. The testability bus can also be used by external test

TABLE 2-1. Economic Impact of Built-in Test

Cost Component Of Total Product Cost	Without Built-in Test ($)	With Built-in Test ($)
Design	384,256	416,345
Programming	412,698	306,500
Fixturing	101,456	25,550
Equipment	753,867	403,668
Test (mfg)	133,540	121,500
Diagnostic (PCB)	308,650	155,340
Diagnostic (system)	619,300	341,400
Parts	301,455	353,342
Service	198,763	54,215
Total Life Cycle Cost	3,213,985	2,177,860
Savings	$1,036,125	

Note: Savings shown cover five-year life cycle of product.

equipment to enhance fault isolation when a defective board is removed after built-in test diagnosis for troubleshooting (or during the initial manufacturing process).

Economic Impact of Built-in Test.   It is important to calculate a total product cost summary for a run of systems, each containing a number of printed circuit boards, with and without built-in test capabilities. Table 2-1 shows such a summary. As shown, very often even as much as a 12 percent extra investment in design cost and parts cost can still result in a 33 percent decrease in overall product life cycle costs. Thus a small investment in testability (a week or less per board to select the necessary control and visibility points and incorporate the testability circuits and built-in test processor on each board) can be turned into a savings of over $1 million over the typical five-year life of a product. Such figures are typical for the costs of and return on investment from enhanced testability.

Standardized I/O Pin Configurations and Ground/Power Pins

Standardization of module, board, and subassembly connection methods simplifies test adapter designs and also reduces the number of different test adapters required for any given system. Items of like size and like connectors should be designed to use the same pins for power and ground inputs and other lines common to multiple items. In addition to power and ground pin standardization, standardization should also be achieved for interlock pins, address, data and control busses, other I/O fields, identifier pins, and control and visibility interface connections.

Minimum Use of Connector Types

As with using standard pin configurations, standardization of connector types reduces the number of types of test adapters required and improves manufacturing and logistics conditions.

Extenders/Cables for Subassemblies and PCBs

In the maintenance or troubleshooting of subassemblies in which PCBs or wire-wrap boards are mounted, it is generally necessary to use extender cards or extender cables. This requirement may also exist when system level ATE is used. The use of extenders can cause additional capacitance, resistance, and inductance to be added to a card input or output. All design efforts should be conducted with these additional test requirements in mind. It is also helpful if the extender cards themselves are designed to facilitate the connection of test equipment to them (e.g., ground terminals, staked pins for ribbon cable terminators, etc.).

Test Points/Test Connectors

Every test point added to a design has the potential for reducing fault simulation and fault isolation time by as much as 50 percent or more per fault, depending upon circuit type and configuration and troubleshooting methodology. Thus, a strategic system decision must be made very early in the design phase as to how (not if!) critical test points will be made physically accessible. Discrete test points provide the least expensive means of nodal data for fault isolation and can be implemented in a variety of ways. A signal can be brought out to a single test jack, for example, as shown in Figure 2-4.

When many test points must be made accessible and physical access is limited, some sort of scan design, shift register, multiplexer, or combination serial/real-time addressable testpoint access scheme should be included in the system architecture.

Nodal points on PCBs should be considered as a valuable source of

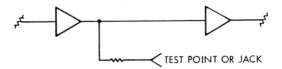

FIGURE 2-4.   Test points/test connectors. The decisions on how test points will be physically implemented needs to be made early in the system design phase so that packaging concepts will facilitate easy access to them.

test information from both an electrical and a mechanical point of view, especially if in-circuit testing is to be employed for board level testing, and these nodes should be made accessible when laying out the board. The test point/test connector method utilized in any design should take into consideration not only the technical requirements of the system but also the test equipment to be employed in manufacturing and maintaining the system.

Test points are most effective when they are unique; that is, each test point should provide test and troubleshooting data that others cannot. This approach reduces the number of points needed and improves logistics considerations and fault isolation resolution.

Visual Indicators

It is recommended that operational status and diagnostic indicators (LEDs, meters, etc.) be provided on PCBs or chassis as part of the original design. These indicators, as shown in Figure 2-5, can be either permanent or pluggable for test purposes only. Besides being a valuable maintenance aid, they can also be used to make a system look more valuable by presenting the built-in test features in a concrete and highly visible manner.

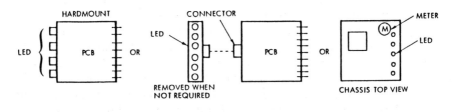

FIGURE 2-5.   Visual indicators. If visual indicators of pass, fail, or error conditions are included in a design, they can make it easy for a technician to identify a defective assembly immediately.

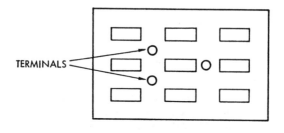

FIGURE 2-6.   Ground points. Making accurate measurements is difficult unless ground termination points are located near to the circuit nodes where critical measurements might be made.

Ground Points

Ground terminals should be provided at various positions on a PCB and/or equipment chassis for use as scope, logic analyzer, in-circuit emulator, and digital multimeter (DMM) grounds in order to avoid using clips on an IC ground pin or having to force using a piece of wire in a connector.

The preferred method is illustrated conceptually in Figure 2-6. Local ground points are especially essential for accurate measurements when dealing with high-frequency (HF), ultrahigh-frequency (UHF), radio frequency (RF), and other types of critical circuits.

Minimizing Adjustable and Select-on-Test Conditions

The number of adjustments required in a design should be minimized. Adjustments add to ambiguity during circuit simulation and take time during testing and troubleshooting. They are also the first line of attack for service personnel who should definitively isolate and repair a defect (rather than compensating for it by tweaking an adjustment). Adjustments are, however, preferred over select-on-test components.

A select-on-test condition, where correct component values are selected on an experimental basis during the testing operation to make the circuit work properly, automatically doubles test time. The component value must be ascertained by the tester, the component must be installed, and the assembly must be tested again to insure that the installation action was correct and did not create other faults on the assembly.

The problem associated with adjustments and select-on-test conditions is compounded when multiple assembly interactions enter the picture or when interdependent adjustments are required. In this day of sophisticated design methods and available components, most adjustments and almost all select-on-test conditions point to substandard design practices.

Digital Feedback Loops

As mentioned in Chapter 1, feedback loops present problems for logic and fault simulation, test pattern generation, and troubleshooting. They are especially troublesome when trying to fault isolate definitively to a specific component on a printed circuit board.

The purpose of fault isolation is to isolate a fault to a replaceable component. In a logic feedback system, it is impossible to definitively fault isolate with the loop closed. And the larger the loop, the more difficult it becomes to troubleshoot and the larger the potentially defective component ambiguity group becomes. Various techniques can be employed to break up feedback loops, and some of these are shown in Figure 2-7.

The physical or mechanical methods are often less than optimally desirable in most systems. Thus, the electrical method is most often preferred. Proper component selection (e.g., a two-input gate instead of a one-input gate) can make the electrical breaking of a feedback loop extremely easy with no impact on circuit performance or reliability (beyond the impact of the extra fan-in connection and its associated resistor). Care should be taken when implementing electrical solutions so that faults are not masked when the feedback loop is broken for fault isolation purposes.

Generic Part Numbers

It is often very difficult to troubleshoot and isolate a problem when circuit information for a specific function is presented on several schematics or drawings. To ease troubleshooting procedures, whether field or factory, it is good practice to put as much circuit information as possible on each relevant schematic. This includes details such as generic component types and values. Company part numbers alone do not provide much data and may necessitate multiple cross-references between documents in order to determine what a part is or what its function is.

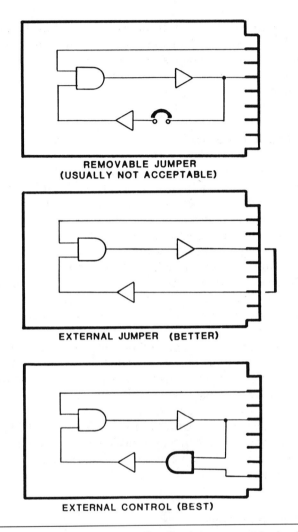

FIGURE 2-7.   Digital feedback loops. Feedback loops must be broken in order to definitively determine which component in the feedback loop is actually causing a fault.

Component Reference Designators

When troubleshooting, a technician is generally using schematics and a top assembly layout to locate various components and test points. Providing reference designators for subsystems, subassemblies, and on printed circuit boards adjacent to critical test points and most (if not all) components can serve to dramatically reduce troubleshooting time.

Timing Diagrams

Timing diagrams should be provided and are extremely helpful for assembly, subassembly, and system test and troubleshooting. Timing diagrams are also used for manuals, training courses, and operations training. If you created (or used) a timing diagram during system design and/or debug, transfer it to test engineering, field service, and technical publication so that they do not have to reinvent it (or come back to bother you for it).

Nodal waveforms are also particularly helpful in conveying information about analog circuit operation. An example of a preferred timing diagram for a digital circuit is shown in Figure 2-8.

The logic, circuit, and fault simulators available on many computer-aided design (CAD), computer-aided engineering (CAE), and computer-aided test (CAT) systems can automatically provide much of the timing and waveform information needed by people in other organizations. Each designer should take advantage of the capabilities of his or her capital equipment and software resources to eliminate duplication of effort in every instance and in every function of the organization.

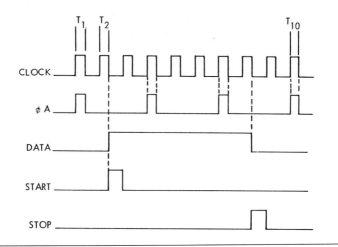

FIGURE 2-8.   Preferred timing diagram example. Timing diagrams used during design debug should be transferred to test engineering so that they do not have to be recreated.

Functional Packaging

Each PCB or subassembly should be designed as a properly partitioned, controllable, and observable functional unit, rather than being split up into bits and pieces of logic or circuits on separate modules or subassemblies. When functions are split, particularly in what appears to be a random fashion, it complicates the test equipment, the test programming effort, and the actual testing and troubleshooting tasks. The functional packaging approach (i.e., having each function complete within a single package) will simplify test generation and test operations.

This approach also aids in diagnosing problems, since the functions missing on other modules or assemblies do not have to be simulated by ATE or BIT resources. It is all right to have multiple complete functions in a single package as well, as long as the guidelines for partitioning, control, and visibility are kept in mind.

Critical Measurements or Adjustments

Production test time standards are often based on general test requirements. Sometimes these estimates are recorded in preliminary documents along with the type of test equipment needed, the number of test stations required, and estimated production costs. Indicate early in the design phase any critical measurements or adjustments that are anticipated and/or highlight critical test areas on the schematic and in all other test requirement documentation. This will reduce possible misunderstandings due to such things as latent multistep initialization sequences, state-of-the-art measurement requirements, noise-sensitive tests, and the like.

3

General Digital
Circuit Guidelines

Electrical design and physical layout of digital printed circuit board assemblies, or programmable logic devices used to shrink previously discrete board level glue logic devices, are often (correctly or incorrectly) influenced by real or perceived physical space or gate count constraints.

Testability concerns are seldom given the attention they deserve during the initial circuit design stages. Yet the ease with which a digital device or printed circuit board can be simulated (both for good circuit operation and for fault coverage), tested, and fault isolated should be one of the design engineer's major concerns.

Employing a few basic testability guidelines during the design phase of a product can result in far-reaching rewards during both the design verification stage and the manufacturing and service lives of the assembly. The purpose of this chapter is to discuss the most common (and sometimes most commonly ignored) and most important general digital circuit testability guidelines. The guidelines that follow, commonly referred to as ad hoc testability guidelines, are equally applicable to programmable logic devices, custom and semicustom devices (ASICs), and printed circuit board assemblies (regardless of the level of integration used to implement each individual design).

The guidelines presented in this chapter are only beginning to be taught to undergraduate and graduate engineering students. Most semiconductor device data books show applications examples that completely ignore the guidelines you are about to read. Most designs, even if they are ostensibly LSI/VLSI based, still contain more than their share of the testability problems that the following (previously SSI/MSI categorized) guidelines will either ameliorate or solve completely.

Whether you design devices, boards, or systems, the following guidelines are critical if your design is to exhibit the basic inherent testability features that are so necessary for high-quality, highly competitive designs.

INITIALIZATION

Initialization means setting a starting position, state, or value. When testing any digital circuit containing sequential (or memory) elements such as counters, shift registers, and flip-flops, it is important that the initial states of each device be immediately controllable in order to insure that all required functions are properly operating. This is true whether the assembly is tested on a bench, in a system, with ATE, or with a built-in test scheme. ATE, in particular, requires that circuits be initialized before testing takes place. Thus it is necessary to provide direct controls for all memory elements undergoing automatic testing.

Counters, shift registers, and other sequential logic are normally self-initializing in a system environment, but are not necessarily self-initializing in their subsystem configurations. *If a circuit or subassembly cannot be initialized, it cannot be tested.* Initialization features are required attributes for predictability and repeatability of logic and fault simulation as well as for testing and accurate fault diagnosis. A lack of immediate initialization causes unknown states to be propagated to other circuit nodes during logic simulation and causes automatic test pattern generation software significant difficulty. It may render it completely unable to cope. When circuits are difficult to initialize, the result is increased software expense, increased test time, restricted test limits, lower than optimum fault coverage, time-consuming troubleshooting procedures, and lower product yields.

Initialization Techniques

There are many methods for providing proper initialization of the various memory elements prevalent in most digital designs. Some require only the use of unused device package or PCB edge connector pins (or IC clip access), while others require the addition of a logic element or extra fan-in input whose only function is to enhance testability. In any case, the circuitry additions are relatively trivial and the results, from a test standpoint, are relatively major.

When selecting components (or CAE library circuit elements) for a new design, select devices that include the capability for direct immediate initialization. Initializable components and elements can be recognized quite easily. They feature an input pin called RESET (or its equivalent). Without the reset feature, a serial initialization sequence will have to be applied to the circuit. If there is direct access to all circuit inputs, that is usually not a problem unless the initialization sequence is either very long or conditional on uncontrollable (via hardware access) internal device states. If an uninitializable circuit is "buried" in the middle of complex logic, it is very definitely a problem.

Basic Initialization Guidelines

Shown in Figure 3-1 are four examples of how to provide for the initialization of the basic memory element, the flip-flop. The methods shown include the following:

- Bringing the SET and RESET (or PRESET and CLEAR) lines to the edge connector is normally required if they are used for system functions (1). Care should be taken, however, to insure that noise will not be injected into the lines shown.
- Providing pull-ups to V_{cc} and bringing out the SET and RESET lines to unused connector pins also allows for good initialization (2) (this is the preferred method).
- If no edge connector pins are available for the additional control points, tie them to V_{cc} through a pull-up resistor for IC clip or bed-of-nails fixture access (3).
- If no edge connector pins are available and the lines are used internal to the UUT, provide IC clip or bed-of-nails fixture access (4), or access via a testability bus interface.

Providing clip or bed-of-nails access implies that the normal states of the initialization lines will be overdriven (or backdriven) by test equipment with high-current pin electronics drivers. This can be detrimental to component life if not used with care. If a testability interface is included in a design, the required reset (or equivalent) lines should be gated such that their states can be changed via the testability interface.

Full initialization, one test pattern after power application, should be used as the design goal for initialization. If multiple patterns are required, go/no-go testing can be done, but faults that prevent initialization are extremely difficult to diagnose.

The configurations in Figure 3-2 should be avoided at all times.

Preferred

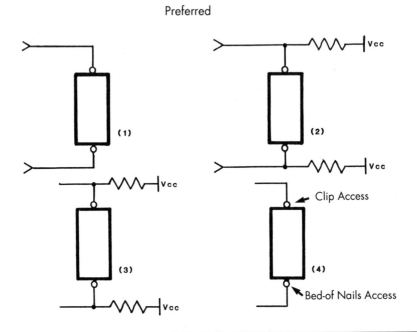

FIGURE 3-1.   Basic initialization techniques. Providing edge connector control, or at least bed-of-nails fixture or IC clip access, to set and reset lines of memory elements is very helpful for initializing the circuit under test.

Non-preferred

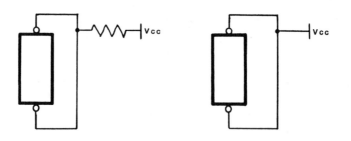

FIGURE 3-2.   Uninitializable circuit configurations. The circuit configurations shown cannot be directly initialized unless the other inputs are connected directly to device or circuit board input pins.

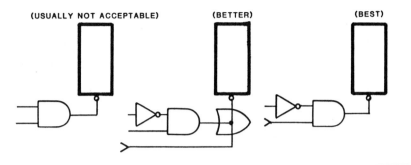

(USUALLY NOT ACCEPTABLE)   (BETTER)   (BEST)

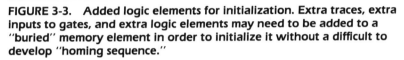

FIGURE 3-3. Added logic elements for initialization. Extra traces, extra inputs to gates, and extra logic elements may need to be added to a "buried" memory element in order to initialize it without a difficult to develop "homing sequence."

Connecting both the SET and RESET lines directly to either the power or ground rails makes it impossible to use them for initialization purposes. Connecting both lines to the same pull-up (or pull-down) resistor results in a race condition when the lines are pulled up (or down), resulting in an unknown state, which makes the situation just as bad.

Added Logic Elements or Functions for Initialization

In certain cases it may be necessary to add a logic element to a PCB or device design to allow for proper initialization as shown in Figure 3-3. Bringing the control of this extra logic element to a device package pin or PCB edge connector pin eliminates the need for an initialization sequence at the device level and for attaching IC clips or using a bed-of-nails fixture at the board level during the production test process, and thus it reduces test time. It may sometimes be possible to avoid adding the extra logic element by implementing a wired logic function instead, but this is the only type of situation in which a wired logic implementation is recommended.

Additional Initialization Guidelines

When multiple memory elements are present in a design, the number of physical control points required for testability can be reduced by using a *master reset* scheme as shown in Figure 3-4. For board level designs, edge connector access is best (1). If in-circuit testing will be performed, the configuration shown should be modified so that alternate packages in long counter chains can be initialized with two master reset lines.

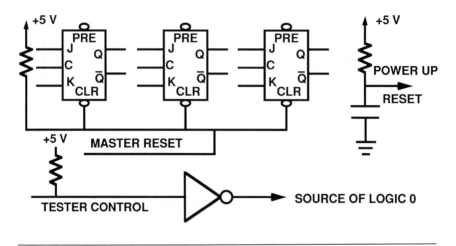

FIGURE 3-4.   Additional initialization guidelines. A single master reset line is often easy to implement. A power-up reset is helpful, and a controllable inverter can be used in place of a hard connection to ground.

A power-up reset can also help from the standpoint of go/no-go testing, but it provides little help in the area of fault diagnosis if the power-up reset circuit fails (2). It is always advisable to have physical control over the power-up reset circuit. The method to use when a logic 0, rather than a logic 1, is needed for normal circuit operation is to use an inverter with its input tied high (through the appropriate pull-up resistor) as a source of a logic 0. This allows an external line or an IC clip to be attached to override the normal logic state and allow for immediate initialization (3).

Connecting unneeded (from a functional design standpoint) initialization points to voltage or ground bus points should be avoided. This same advice also applies to other unused (from a functional standpoint) inputs to devices on a PCB. They should be connected to the appropriate logic level via a pull-up or pull-down resistor or through an appropriate logic element.

Initialization Examples

The examples in Figure 3-5 show a frequently used divide-by-4 counter which generates internal clock pulses and runs continuously as long as the INHIBIT line remains low. However, when the INHIBIT line goes high, the counter continues to run until it reaches a count zero state; thus it is self-initializing—when it works.

Two factors limit the suitability of this type of circuit for automatic

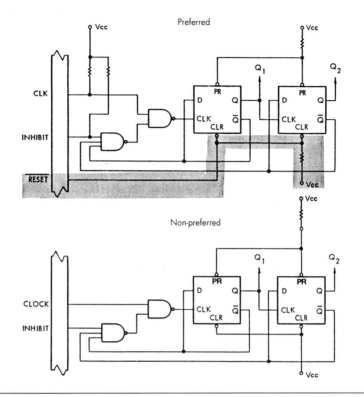

Preferred

Non-preferred

FIGURE 3-5. Circuit initialization examples. The circuit with the reset line can be directly initialized from the card edge connector. The circuit without the reset line requires manual intervention with an IC clip or a bed-of-nails fixture.

testing. First, the circuit requires one to three clock pulses for logic initialization after power-on. As the number of divide-by-2 networks increases, the number of required initialization clocks increases exponentially. Second, the circuit may never be initialized if a fault occurs during the initialization procedure. In such a feedback network, faults are difficult to find because they are often not repeatable. By designing the circuit as shown in the previous example, automatic testing can be achieved. In the system design, the RESET line can remain unconnected in the back-panel wiring if it is not needed for system functions. If it is not possible to bring the RESET line(s) out to an edge connector, physical access via an IC clip or bed-of-nails fixture should be provided.

Self-Resetting Logic

Self-resetting logic (see Figure 3-6) presents a testability problem in that the flip-flop may reset itself so quickly after power is applied to the

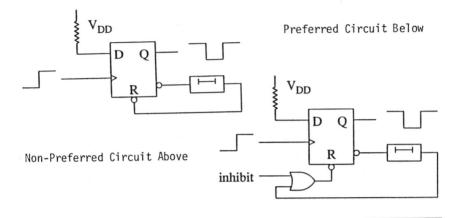

FIGURE 3-6.   Self-resetting logic. Circuits with self-resetting logic are feedback loops that need to be broken in order for automatic test equipment to control them and detect the results of their functions.

circuit that the test equipment cannot accurately or definitively observe the set logic value. This is especially true if the delay line time is extremely short (e.g., in the nanoseconds range).

Rather than using the uncontrollable version of the circuit shown in the left-hand side of Figure 3-6, use the controllable version shown in the right-hand side of the drawing so that the self-resetting function can be inhibited for test purposes.

ASYNCHRONOUS CIRCUITS AND ONE-SHOTS

Asynchronous circuits of any type should be avoided in digital circuit designs wherever possible. The time dependency of their operation makes testing difficult because they can be sensitive to tester signal skew or require constraints on the timing of input logic state changes. Asynchronous circuits also exhibit nondeterministic behavior when fault simulation is performed, and good design of asynchronous circuits often requires redundant logic which can reduce fault coverage by creating undetectable faults. Asynchronous circuits are usually incompatible with structured design for testability techniques (e.g., scan design) and with any built-in self-test hardware surrounding them. Ripple counters and mixed synchronous/asynchronous circuits fall into the category of asynchronous circuits, but the most often seen example is the monostable multivibrator, or one-shot.

Normally, one-shots designed into circuitry present an asynchronous characteristic in their outputs and do not easily lend themselves to automatic testing. Depending on the type of ATE used, testing of the one-shot and the other circuitry may need to be implemented as separate steps in the test procedure.

Two methods for providing input stimulus control and output visibility are illustrated in Figure 3-7. The upper portion illustrates using a mechanical jumper at the edge connector to isolate two cascaded one-shots. In the lower portion, a select gate is added to isolate the one-shot so that the output circuit can be externally stimulated. Provision has also been made for directly monitoring the one-shot's output pulse width when it is stimulated through the OR gate.

The input and output of a one-shot can be isolated from the other circuitry with jumpers at the edge connector or by means of a select gate. To test the one-shot itself, the output of the select gate can be brought out to the edge connector, or the one-shot output itself can be brought out as a test point as shown. If the circuit does not have outputs at the edge connector, the output to the one-shot should be made available via IC clip or bed-of-nails access.

If one-shots must be used, and if they must be cascaded, the approach to testing them should follow the example in Figure 3-8. A typical problem in testing cascaded one-shots occurs when $T_A >> T_B$, making parametric measurement with an automatic waveform analyzer impossible.The modification shown in the circuitry on the left makes it possible to test one-shots automatically and reliably with a high degree of accuracy.

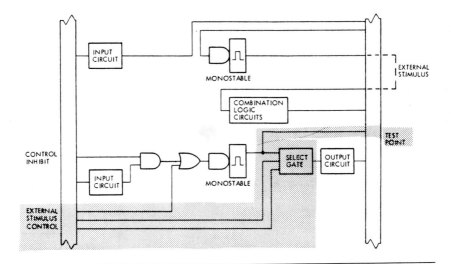

FIGURE 3-7.   Control and visibility for one-shots. One-shots, or monostable multivibrators, should be separately isolatable from the logic they drive and separately measurable using either electronic or mechanical methods.

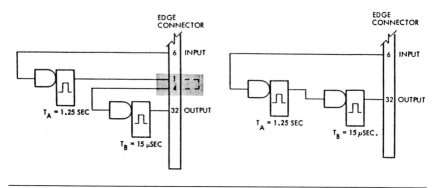

FIGURE 3-8. Cascaded one-shots example. If one-shots must be used, and they must be connected in series, some means to disconnect them should be employed, especially where large differences in pulse width are present.

With either the select gate or the physical jumper approach implemented, each individual one-shot circuit can be tested as an entity. Without one of the modifications, it is very difficult to measure, for example, a 15-microsecond-width pulse width that occurs only once every 1.25 seconds.

One-shots also interfere with testing in general and with testing of adjacent (or following) logic in several ways. First, the pulse width of the one-shot may be so short that it cannot be detected with normal functional digital test patterns. An external capacitor may need to be added to the circuit as shown in Figure 3-9 (left side) in order to lengthen the pulse width to a detectable value. Or the time constant of the one-shot may be so long as to make it very time consuming to test the logic being driven by the one-shot. An external resistor may need to be added to the circuit as shown in Figure 3-9 (right side) in order to shorten the pulse width to an acceptable value.

Because one-shots cause so much trouble in testing, it is best to replace them with *synchronous logic* implementations whenever possible.

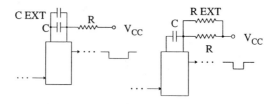

FIGURE 3-9. Modifying one-shot time constants. The time constants that control one-shot pulse widths can be lengthened or shortened by adding capacitors or resistors in the test fixture.

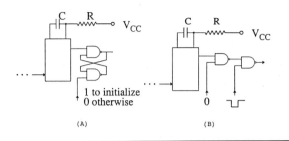

FIGURE 3-10.   Blocking and gating around one-shots. One-shots can be isolated from other logic through the addition of two extra gates.

Testing is often performed at speeds different from normal circuit operating parameters for a variety of reasons. One-shots present problems in this case since the one-shot pulses may not end at the proper times. Further, one-shot outputs can jitter and be falsely triggered by other signals on a board. Thus the one-shot should be isolatable as shown in Figure 3-10(A.) And, rather than depending upon the one-shot to drive the following logic, it can be blocked and bypassed by using the logic-gating circuit as shown in Figure 3-10(B).

INTERFACES

Circuits designed with multiple logic levels may require special considerations. Many ATE systems offer multifamily programmable driver outputs and sensor thresholds, but these sophisticated systems are more expensive than single-family testers. If a single-family tester is all that is available for testing a specific design, level shifters, buffers, and other circuitry may have to be added to test adapters in order to test the additional logic levels. To lower these interface costs, make all I/O levels compatible for any given design, if that is at all possible and practical. If mixed logic levels must be used, try to sandwich the non-standard levels between standard levels so that the standard logic is at the board edges. The non-preferred and preferred methods are illustrated in Figure 3-11.

BUILT-IN TEST DIAGNOSTICS

It is important to remember that testing, including built-in test hardware- and software-implemented testing, requires both go/no-go testing (as a minimum) and the ability to diagnose where in the circuitry

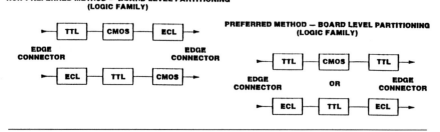

FIGURE 3-11.   Mixed logic levels. Mixed logic levels require either more sophisticated automatic test equipment pin electronics or active electronic circuitry in the test fixture.

a fault has occurred. In the design of a BIT circuit where many signals may be ORed together to provide a single fault indication for functional purposes, it is also possible to provide diagnostic information if a little extra circuitry is implemented.

In Figure 3-12, extra latches are implemented, as shown on the right-hand side, so that if a fault occurs in any of the circuits its occurrence is latched to facilitate troubleshooting. Both circuits function identically from the user interface standpoint (i.e., a failure in circuit segments A, B, C, or D activates the fault indicator), but the circuit on the right provides far better diagnostic resolution.

The outputs from the extra latches can be brought out to circuit edge connector pins for external ATE testing and troubleshooting or con-

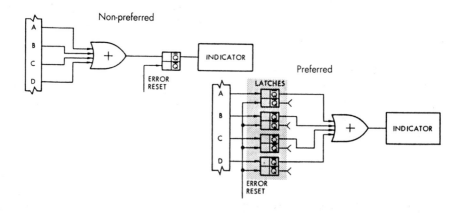

FIGURE 3-12.   BIT enhanced for diagnostics. Knowing where a fault is, rather than just that a fault exists, is important for fault isolation. The diagnostic version of built-in test requires a few more gates but shortens troubleshooting time.

nected via a multiplexer to an on-board built-in test processor or via a testability circuit to a testability bus connected to an off-board built-in test processor. In any case, fault isolation, which is a big subset of testability in general, has been considerably enhanced.

FEEDBACK LOOPS

Feedback loops occur in most moderately complex circuits and they complicate the design verification, logic and fault simulation, and automatic test generation processes. Complex loops, especially those containing memory elements, will often also have to be separated into smaller segments for fault isolation using ATE backtracing algorithms. Generally, all of the nodes in a loop do not change state together. In some loops, troubleshooting difficulty arises when errors propagated through the loop are fed back to the beginning as well as to the edge of the board where they are first detected.

Feedback loops should thus be controllable. The preferred and unpreferred configurations for the generalized feedback loop case are shown in Figure 3-13. Segmenting a loop may or may not require additional logic devices. A trace routed to the edge of the board, where a jumper is used in the normal system operation, for example, helps to simplify testing. Another approach would be to use a gate in place of one

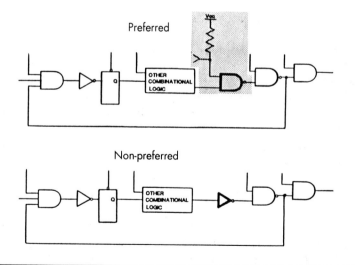

FIGURE 3-13.   General feedback loop example. Whenever a feedback loop exists, it is often possible to provide the means to break it by using an $N+1$ input gate (i.e., a two-input gate instead of a one-input gate).

Gates and Multiplexers   Three-state Nodes

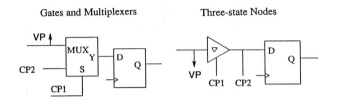

FIGURE 3-14. Additional feedback loop control methods. Multiplexers or tristateable gates can be inserted in one-shot paths to provide for control and visibility to ease the test generation and troubleshooting tasks.

of the inverters in the loop (as illustrated in Figure 3-13) and wire the additional input through a resistor to V_{cc} as illustrated. This input can then be driven low by the tester to interrupt the loop. Controlled access within feedback loops helps in identifying fault conditions, especially when a loop contains many ICs.

Feedback loops can also be broken by inserting extra logic in them to provide for both visibility and control. Or a single-function circuit can be replaced with a more powerful multiple-function circuit to accomplish the same thing without inserting extra propagation delays in the feedback loop. Two additional methods for breaking feedback loops in this way are shown in Figure 3-14 (where VP stands for visibility point and CP stands for control point).

OSCILLATORS AND CLOCKS

One of the most common testability problems in logic design is a free-running oscillator buried within the logic. A buried oscillator, or clock, is one which is not exposed to a board edge and is not controllable by the testing device. The tester must then establish its own time reference and maintain synchronism. If the internal clock speed of the circuit is faster than that of the tester, then the unit under test logic can sequence through several states during one tester clock cycle, which makes it impossible for the UUT outputs to be verified.

There are several ways to free a buried oscillator to prevent such an occurrence. The following two are shown in Figure 3-15.

1. Provide for isolation of the oscillator. Partition the circuit so that the output is brought to the board edge. Then provide an external oscillator input to the logic circuitry that can be jumpered at the board edge for normal operation or used as a direct input from the tester. Locate the oscillator circuit near the board edge connector to allow short runs and minimize crosstalk.

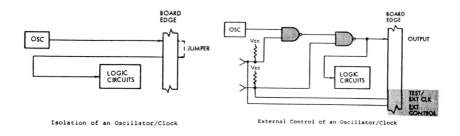

Isolation of an Oscillator/Clock   External Control of an Oscillator/Clock

FIGURE 3-15.   Clock control methods. Either mechanical or electrical means can be used to block a free-running clock or oscillator and substitute a clock from the automatic test equipment.

2.   Provide for external control of an oscillator. Include the means to disable the UUT oscillator and allow an external tester clock to be applied. In normal operation the test/external clock and external inputs are open. When testing, the tester drives the external input low while providing its own clock to the test/ external clock input. The output of the circuit is tied to the logic, and it is desirable to bring the output to the board edge so that clock parameters may be verified.

FAN-IN AND FAN-OUT CONSIDERATIONS

Some digital circuit configurations result in many logic elements having their functions combined into a single signal term, which is the result of a very complex logic function or equation. The point where all of the functions converge to be combined is termed a *fan-in point*. A high logical fan-in structure (see Figure 3-16) is difficult to control to one of its two output values. It also makes it difficult to observe the effect of the individual input logic states on the logic feeding the fan-in point. Extra inputs at intermediate stages (or replacing one-input gates with two-input gates) makes it much easier to test structures like the one illustrated.

A high *fan-out point* is a single point that drives many succeeding logic devices in parallel. Device fan-out should always be limited to at least one less than the maximum allowable number of loads. Inside an integrated circuit (such as a gate array or custom IC), limiting fan-out to less than the maximum allows for the addition of a visibility point (or observation point) for testability without violating the loading rules for the IC logic element.

At the board level, limiting fan-out lets an in-circuit tester, or the guided probe typically associated with troubleshooting on a functional tester, be used without causing degradation of the circuit performance

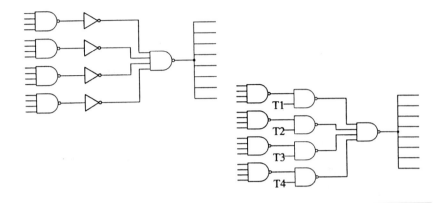

FIGURE 3-16. Fan-in and fan-out structures. Visibility is needed at fan-in points like the output of the right-hand NAND gate, and control of the inputs to the fan-in point that fans out to other circuit nodes is also desirable.

due to excessive loading. If maximum fan-out is used totally for functionality, the connection of a tester or probe may actually cause the circuit under test to cease to operate.

BUSSED LOGIC

Systems designed with interconnecting busses may present major circuit board testing problems, especially if the bus is contained entirely on the board. It then becomes a problem similar to testing large fan-in and fan-out logic implementations. It is best if all busses are physically accessible to an external ATE system (either directly or through some testability interface) and to any on-board or in-system built-in test resources.

However, with bidirectional busses, where separate receivers and drivers are used, isolation is easier if some of the devices are connected to the bus through jumpers. Removing a jumper (or lifting one leg of the jumper), while not the best solution, is far easier than unsoldering a multipin device. A better approach is to have all tristate ENABLE lines brought out to an edge connector (either directly or through some testability interface).

When an input bus and an output bus are to be connected together, it is best to make the connection in the edge connector rather than permanently on the circuit board. The separation could also be made using IC sockets with jumper plugs, or PCB switches, at a somewhat greater material cost.

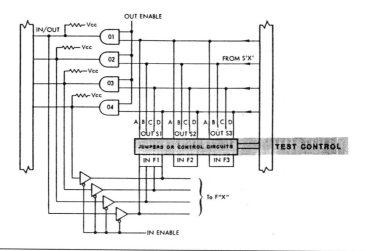

FIGURE 3-17.   Test control for bussed logic. Providing either electrical or mechanical control of bussed logic allows faults to be isolated more easily.

In CMOS-driven busses, using pull-up resistors on the lines will help in troubleshooting when failures occur. In the circuit shown in Figure 3-17, a fault on any of the lines on the right-hand side can be caused by any (or all) of the inputs. The control circuits (or jumpers) allow each portion of the circuitry to be disabled in order to aid in accurate fault isolation.

BUFFERS

The purpose of adding buffers is to prevent stray noise from entering a circuit where memory elements might be arbitrarily set or reset by externally induced noise and/or to allow the ATE to source or sink current to or from a circuit that may otherwise be at its maximum fan-in/fan-out limits. The example shown in Figure 3-18 demonstrates both the preferred and unpreferred circuit designs, where the inclusion of a buffer eliminates potential testability problems.

Buffers can also be added to intermediate points within the circuitry to minimize loading and to decouple test logic from normal operating logic. In the upper portion of Figure 3-19, an inverter is added to the connection between IC1's output and IC2's input to allow the external visibility point to be observed without affecting normal circuit operation. In the lower portion of Figure 3-19, an extra transistor in an IC is used so that any loading effects from the visibility point only occur when the transistor is enabled via the T input.

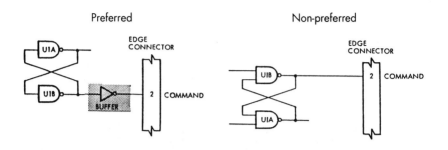

FIGURE 3-18.   Buffer added to memory element. Memory elements constructed from simple gates are susceptible to unwanted states due to noise being driven back on their outputs and should thus be buffered.

- Buffer observation points to minimize loading.

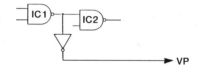

- Decouple test logic from normal operational logic.

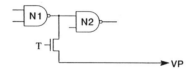

FIGURE 3-19.   Buffers to minimize performance effects. Providing visibility can sometimes affect circuit performance. Buffers or transfer gates can be used to minimize these effects.

Buffers should also be provided for edge-sensitive external inputs. Testers may not always provide the proper rise and fall times compared to the actual system circuitry that the circuit under test will see in its end use. If the buffers cannot be placed on a board, they may have to be placed in the test fixture, which, while certainly feasible, is more expensive. Buffers for edge-sensitive external inputs are especially important for external clock inputs, particularly if those clock inputs are distributed on-board directly to many sections of the board's circuitry.

VISIBILITY POINTS

Visibility points are one of the easiest ways to provide drastically improved testability. Whether the visibility points are brought out to multiple discrete physical connections or registered or multiplexed to a testability connector, they significantly ease the fault simulation, test generation, and troubleshooting tasks. Once a UUT fails a test, the fault must be located. ATE uses test points to help isolate faults automatically and the inclusion of visibility points often makes it possible to perform automatic fault isolation with improved ease and accuracy.

The first guideline for visibility points is this: If you had an instrument connected to a circuit node (or added an extra pseudo-primary output to the CAE design file for the circuit) to make it easier to verify the design and/or debug the prototype circuit, that point should be made a permanently accessible test point for the manufacturing test and maintenance people.

The next guideline, illustrated in Figure 3-20, is to use the "cut it in half" method. Simply place the first test point in the logical center of the circuit. Then place additional test points in the logical centers of the new partitions thus created. Use as many test points as can be reasonably implemented in each design since each visibility point has the potential for reducing troubleshooting costs by 50 percent or more per fault.

Another good place to add visibility points is at the output of logic driving display or other output devices. In Figure 3-21, it can be seen that having to verify circuit performance by having a human verify multiple light-emitting diode (LED) on/off sequences during testing is time-consuming and error-prone. To reduce the cost and increase the quality and repeatability of test operations, advantage can be taken of ATE or BIT by measuring logic levels via the added test points and then performing one operator visual test with all LEDs lit.

Test points are a valuable resource, and their locations should be intelligently selected. An example of wasted visibility points is shown

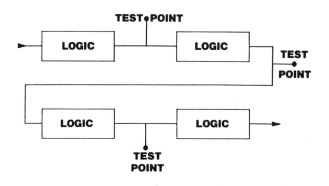

FIGURE 3-20.   Visibility points between logic blocks. Test points placed between logic blocks reduce the amount of probing required to isolate faults. They also make design verification and test generation easier.

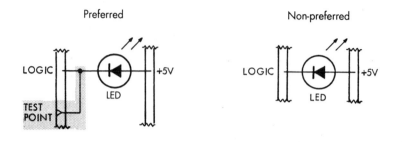

FIGURE 3-21.   Test points at interfaces. Electrical access to the logic driving visual indicators allows the function of the logic to be verified without requiring human observation of the visual indicator.

in Figure 3-22. The dashed lines are not necessary because they all come from the device labeled U1. If any output from U1 is faulty, we would replace U1. This can be determined from the U1D output visibility test point. Were each of the gates shown physically contained in separate packages, however, it would be helpful to have some of the additional visibility points.

As mentioned, the general guideline for visibility points is to add them where they will divide a circuit approximately in half. The result is a significant reduction in the time and cost for all design verification and testing-related tasks.

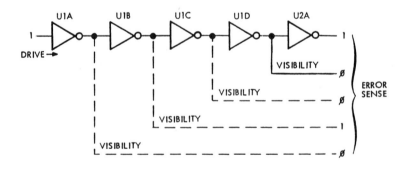

FIGURE 3-22.   Intelligent visibility point placement. Visibility points placed between device packages or other separate modules are much more valuable than test points placed at intermediate stages in a single package.

PARTITIONING FUNCTIONS INTO LOGICALLY SEPARABLE UNITS

Functions should be designed as logically complete units rather than having part of a function on one assembly and the remainder on another assembly. This approach simplifies testing and troubleshooting and facilitates BITE design because fault isolation can be achieved much more readily. Modular design of each function also simplifies testing by reducing the amount of input/output simulation required to test that function. Preferred and unpreferred methods for partitioning a complete circuit contained on the same PCB are illustrated in Figure 3-23. The partitioned version can be tested with 15 test vectors (as opposed to 4,440 for the untestable version). With more complex circuitry, the advantages of partitioning are even more apparent.

WIRED OR/AND FUNCTIONS

It is difficult to debug circuitry that is OR/AND wired. Many technicians use two relatively destructive troubleshooting methods with wired-OR type circuitry: cutting the legs of suspected faulty IC outputs connected to the node until the fault is revealed (used on the component side of the board for boards built with through-hole technology), or cutting the

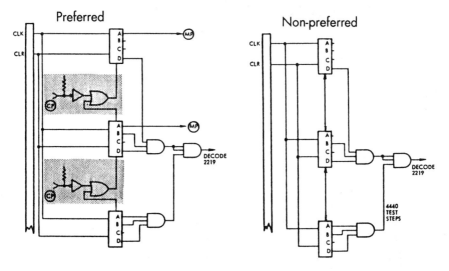

FIGURE 3-23. Circuit partitioning. When circuits, particularly those with counter chains, are partitioned, the number of test vectors required to verify their functions can be drastically reduced.

Preferred   Non-preferred

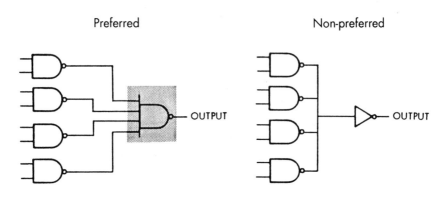

FIGURE 3-24.   Wired-OR/AND circuits. While wire-OR/AND circuits may save
a gate or two, they may require replacement of multiple device packages in
the event of a failure at the board level.

traces between suspected faulty outputs for the same purpose (used on the solder side of the board and with surface mount technology). Alternatively, suspect ICs may have to be physically removed for troubleshooting. This may have been inexpensive in the past, but now, with ASICs and expensive VLSI devices, it is ridiculously wasteful.

Avoid wired-OR and wired-AND connections and use an extra gate if necessary (see Figure 3-24) to enhance the testability and maintainability of the unit under test. When outputs are wired together, any short on the node makes it almost impossible to locate the defective gate.

If it is not possible to eliminate the wired-OR/AND circuit, at least try to put all of the gates associated with it in the same physical device package. That way if there is a failure on the wired node, the fault can be immediately attributed to a specific package instead of to one of several device packages.

COUNTERS AND SHIFT REGISTERS

Several problems arise where counters or shift registers are used, especially if long sequences or chains are implemented. One problem often encountered is lack of initialization; another is the inability to directly control (via the ATE) the data input. Many of the resultant test patterns are thus long and complex, which complicates the test programs and the verification of good circuits.

Another problem involves the inability to control the clock input(s) to a counter or shift register. These clocks are frequently derived from complex on-board logic functions. This also complicates the testing process and the writing of test programs. When on-board clocks are too fast for specific ATE, other techniques must be implemented to test these circuits. Cascaded counters or shift registers, which create long chains, can add to the problem of complex data patterns and test programs. These long chains increase the test time and complexity of the test. There should be no more than four counter stages or chains without a breakpoint.

Figure 3-25 illustrates breaking up a counter chain into the recommended maximum number of stages. To test the circuit shown, it is only necessary to reset the circuit to all 0's, count to 11111111 (15 clock cycles), and then count to 11111110 (14 clock cycles). Thus the test can be completed in only 30 clock cycles. Without the added gates for testability, that same test would take 17 times as many (or 511) clock cycles. Consider that impact when multiple counter stages are even further cascaded.

The testability solutions are relatively simple, but they may require additional I/O pins and/or the addition of logic functions for implementation. The illustration in Figure 3-26 presents some of the simple techniques for partitioning and adding control to a complex circuit containing multiple counter chains. The counter/shift register has been partitioned and designed to be synchronous with a common controllable clock, and each section can be separately stimulated. Only two cascaded registers are shown in the example, but the circuit shown is representative of several sometimes larger configurations.

As can be seen, the tester can now inhibit the on-board clock and supply a synchronized clock of its own. This allows the tester to run at its own speed and control the operation of the register. The tester can also inhibit the on-board data and supply its own data to each of the registers. This may or may not be a desirable feature, depending on the particular test circumstances. The diagram also shows that a test point

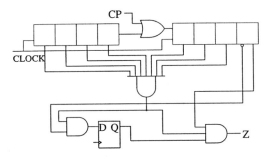

FIGURE 3-25. Partitioning counter chains. A good guideline is to provide intermediate control of counter chains every four stages. This reduces the number of test vectors required to verify correct operation of the counter chain.

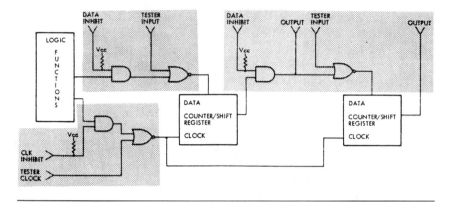

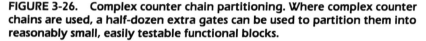

FIGURE 3-26.   Complex counter chain partitioning. Where complex counter chains are used, a half-dozen extra gates can be used to partition them into reasonably small, easily testable functional blocks.

has been added at or near the output of each register and the chain between registers has been broken to allow each register to be tested individually, with data provided either by the tester or from the previous stage. These are all designer options. It may be possible in some cases to combine the tester input control point with the output test point and thus simplify the circuit even further.

Implementing such designs will allow circuits containing counters or shift registers to be readily tested. The tester will also be able to synchronize and control the test sequence rather than have to accept circuit control of the test. In the simple example of Figure 3-26, the portions added for testability appear to be major. In practice, however, the additions tend to be a half dozen gates on a 100-IC board whose total gate count may exceed 500,000. That represents an increase in gate count of about 0.0012 percent and a reduction in test cost, for every board, of as much as 50 percent. Sometimes the needed extra gates are left over as spares just waiting to be used. At other times, it may take an extra IC or two or compressing some other glue logic into an ASIC, in order to make space for testability. In any event, implementing most of the guidelines in this chapter will typically only take up from 1 to 3 percent of board space.

ADDITIONAL GENERAL DIGITAL BOARD GUIDELINES

There are a few additional guidelines that apply to digital printed circuit boards and the devices that they are designed with. These additional guidelines deal with hardware partitioning, delay-dependent logic, and logical redundancy.

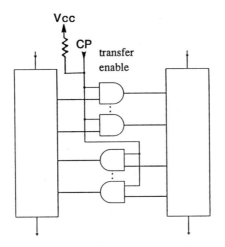

FIGURE 3-27.   Partitioning via hardware. Extra gates, or N+1 input gates, placed between major logic blocks, can provide electrical partitioning for test purposes.

Hardware Partitioning

Circuits may be partitioned into smaller functional blocks through the technique of *degating*. This technique is illustrated in Figure 3-27, where gates are inserted between the blocks to allow them to be totally isolated from each other for test purposes. When the TRANSFER EN-ABLE line is set to the logic 1 state, the circuit operates normally.

Delay-Dependent Logic

Delay-dependent logic, illustrated in Figure 3-28, is often used to gener-ate pulses on boards in place of (or in addition to) one-shots. Delay-dependent logic causes testability problems, however, and its use should be avoided. An automatic test pattern generation program works in the logic domain and thus views the example circuit shown as redun-

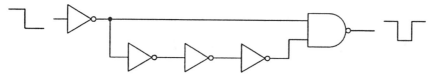

FIGURE 3-28.   Delay-dependent logic. Delay-dependent logic configurations should be avoided because they create undetectable faults and their pulse widths vary, depending on device vendor and operating temperature.

dant combinatorial logic. A fault simulator, even an event-driven one, may find that this circuit configuration creates undetectable faults, thus reducing the achievable fault coverage figure.

Logical Redundancy

Redundant circuits tend to "hide faults" (i.e., to create undetectable faults), and their use should be avoided. Figure 3-29 shows two typical circuits that exhibit logical redundancy. When you see circuits like this, they need testability improvement. If their use cannot be avoided, they should be redesigned with an extra input or an extra gate so that the redundancy can be disabled for test purposes.

Error detection and correction circuits also fall into the category of redundant circuits in that their purpose is also to hide faults. Whenever error correction circuits are used, it must be possible to *disable error correction* for test purposes. It must also be possible to induce errors into the circuit. Otherwise it may not be possible to test all of the circuitry that performs the error correction task.

In parallel hardware computer architectures, it must be possible to turn off one side of the hardware so that the other side can be tested. Then the tested side is turned off, and the other side tested. No other method insures that both halves of the circuit are actually working or allows accurate diagnosis of faults.

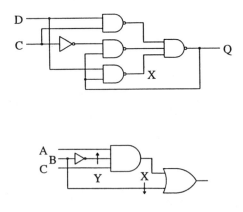

FIGURE 3-29.   Redundant logic examples. Redundant logic hides faults. In test, the objective is to detect faults. Thus redundant logic should be avoided or made controllable so that functions can be tested independently.

GUIDELINES FOR PROGRAMMABLE LOGIC DEVICES

The use of programmable logic devices in place of standard glue logic on digital circuit boards is widespread. These devices, while very powerful from the design standpoint, can create real problems in terms of both test pattern generation and in-circuit board testing. Implementation of the guidelines that follow will considerably ease the testing problems where programmable devices are employed.

Output Enable Control

Allow the ATE full control of the OUTPUT ENABLE pin for registered programmable array logics (PALs). Provide a separate pull-down resistor (or other source of a logic 0) for each PAL. If two or more PALs use a common enabling signal, buffer the signal to each PAL to allow ATE control. This is especially important if the outputs of one PAL are inputs to another PAL and these PALs use a common enable. Do not tie the OUTPUT ENABLE pin to ground at board level.

Initialization

When possible, provide a single pin that will initialize all registers to a stable state when this pin is active. This pin may be tied to the board reset signal or to a separate pull-up or pull-down resistor as appropriate. If an input pin is not available, the designer must provide the test engineer with a set of patterns that will bring the PAL to a stable state when that set of patterns is applied to the inputs. The number of patterns should be as few as possible and must always result in the same final state.

Tristate Control

Provide an input pin that will, when active, tristate all nonregistered outputs. If this pin is not provided, the PAL will have to be backdriven to the desired state when devices attached to the output pins are being tested by the ATE. Since in some designs there is internal feedback, backdriving outputs can cause instability on the outputs, resulting in glitches and, in some cases, oscillation.

If it is not possible to provide an input to tristate the outputs, provide an input pin that, when active, will cause all outputs to be high.

No high output should depend on feedback terms, because the tester may backdrive these pins low. If a pin is not available, the designer must provide the test engineer with a input pattern that will set the outputs high and disable feedback.

Note that this also applies to the nonregistered outputs of the R PALs. This step is important because some PALs cannot be backdriven to the 1 state. Some PALs have output buffers that have noise feedback paths that will cause oscillation.

Internal PAL Design Guidelines

Just as certain circuit configurations present problems at the board level, certain circuit configurations usually found internal to programmable logic device designs cause similar problems (usually with similar circuits). This section deals with the most common problems with these devices and their solutions.

On-Chip Oscillators.   It is quite easy to create oscillators with programmable logic devices (PLDs). Most PLD testers, however, cannot deal very well with verifying the performance of such oscillators. Just as clocks need to be controlled at the board level, on-chip PLD clocks should also be made controllable. Figure 3-30 illustrates both the non-preferred and preferred methods for implementing on-chip oscillators.

Pulse Generators.   Pulse generators (or glitch makers, as they are sometimes called) made up of circuitry which takes advantage of gate delays (e.g., delay-dependent logic) cause problems for automatic test pattern generation not only for the pulse generator circuitry but also for the circuitry fed by it. The non-preferred circuit and a method to stabilize it are shown in Figure 3-31.

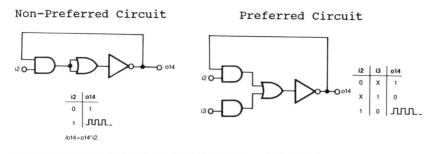

Non-Preferred Circuit   Preferred Circuit

FIGURE 3-30.   PLD on-chip oscillators. Extra input pins should be provided on programmable logic devices so that on-chip oscillators may be disabled and an external clock supplied.

Non-Preferred Circuit   Preferred Circuit

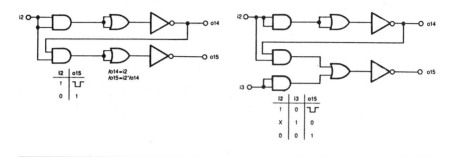

FIGURE 3-31.   PLD pulse generator example. Just as on-board one-shots present problems, on-chip pulse generators do so also. An extra input control allows the output state of the pulse generator to be set to a logic 1 or 0.

Non-Preferred Circuit   Preferred Circuit

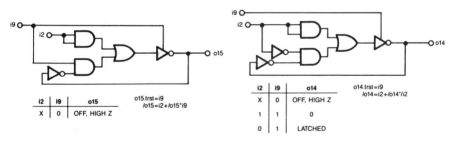

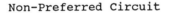

FIGURE 3-32.   Feedback loops with tristate outputs. Separate control of the tristate output where a feedback loop has been implemented is highly desirable.

Memory Elements.   When memory elements are created using the feedback around a tristate type of combinatorial PLD output, special testing problems are introduced. This is especially true when it is desired to tristate the PLD without upsetting its internal states. The unpreferred and preferred methods for implementing memory elements which include a tristate driver in the feedback path are illustrated in Figure 3-32.

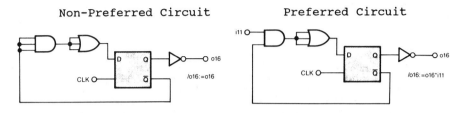

FIGURE 3-33.   Flip-flop initialization. A directly controllable input to internal flip-flops allows for immediate initialization and prevents unknown states from propagating throughout the PLD.

Initialization of Flip-Flops.   Uninitializable flip-flops within PLDs cause simulation, test generation, and test problems at the component and board levels. All memory elements should be initializable via a device input pin. Figure 3-33 illustrates both the non-preferred and preferred methods for implementing flip-flops on PLDs.

4

General Analog Guidelines

Despite the trend toward digital processing in electronic systems, a significant portion of the electronics will normally still contain some analog circuits. Analog circuits are typically characterized as being circuits wherein the signals are in the form of continuous variations in amplitude, phase, frequency, or waveform. The common denominator in such circuits is the circuit loading and signal degradation which inevitably occurs when analog signals are transmitted between the unit under test (UUT) and the test equipment.

Analog circuits can also present challenges to implementing built-in test in that most built-in test implementations are digital in nature. Thus comparators and/or analog-to-digital and digital-to-analog converters (ADCs and DACs, respectively) may be required on occasion for complete built-in test or even adequate board level inherent testability.

This chapter describes guidelines that will make analog circuits in general more inherently testable and thus easier to write test programs for, easier to test, and easier to troubleshoot.

Analog and RF circuits require more specialized treatment than other circuits. Special test setups are often required because there are more variables involved than in the digital cases where one must only verify a specific pattern/time response to a specific set of inputs. Analog circuit designers must consider variable amplitudes, nonlinearities, phase relations, and impedance matching, among other factors. The most important point to keep in mind is that the end product must not only work in conjunction with other elements of a system, but it must also be capable of being tested in an efficient, economical, and timely fashion in the factory and field.

GENERAL ANALOG TESTABILITY GUIDELINES

The following general analog guidelines apply to virtually all analog, or mixed analog and digital, circuits. Like the digital guidelines, they are aimed at helping you to achieve partitioning and control of, and visibility to, the critical internal nodes of the unit under test.

Adjustments

Minimize the use of adjustments requiring trim-pots, trimmer capacitors, and variable inductors. These components extend test times and require operator interaction during testing and verification, not only for go/no-go performance but also for range adjustments and calibration. Also, during operation, adjustable components are sensitive to vibration, drift, and tweaking, all of which may lead to system downtime. Interactive adjustments on different assemblies should also be minimized.

In place of adjustments, design the circuit with components of appropriate precision and rated at their end-of-life tolerance for worst-case environments. An even better approach, particularly in mixed analog and digital systems, is to utilize on-board references, programmable adjustments, and self-calibration with digital techniques.

Relays

Minimize the use of relays and use solid-state components such as field-effect transistor (FET) switches, analog multiplexers, and solid-state power control devices. Relays, being electromechanical devices, are inherently less reliable than solid-state components. Where relays must be used, such as in high-voltage, high-current, and/or high-power applications, provide the appropriate snubbing networks for contact and coil driver protection.

Replacing electromechanical devices with solid-state devices can also reduce board and system sizes, weights, and power consumption in a design, thus leaving more room for additional testability and built-in test features. Such actions also tend to reduce the number of discrete components needed on each board. That benefit reduces parts and assembly labor cost and the size of any ambiguity groups in the event of a fault.

Feedback Loops

Analog circuits, just like digital circuits, typically contain feedback loops that can make fault isolation very ambiguous and difficult. Automatic gain control (AGC), automatic frequency control (AFC), self-zeroing, and phase-locked loop circuits tend to be the most prevalent types of feedback loops in analog designs, and should be able to be broken during test and debug operations.

The principles of and the reasons for breaking analog feedback loops are exactly the same as for digital circuits. Only the physical implementation techniques are different. Instead of an extra input to a gate, or a gate with a larger fan-in, a FET switch or analog multiplexer must typically be used. Sometimes it is necessary to resort to physical jumpers or switches for opening feedback loops, but solid-state methods that can be controlled by ATE or BIT circuitry are preferred.

Open-loop testing is often preferred for preliminary circuit tests and for fault isolation. Final tests should include a thorough exercise of closed loops.

Analog Signal Interfaces

It is very difficult to supply very low level or very high frequency stimulus signals to analog circuits in an ATE environment. It is also very difficult to measure very small signals or signals produced by very high impedance or high-frequency (or both) circuits in that environment. Thus, designs should be partitioned in such a way as to make the signal interfaces at board edge connectors compatible with commercially available ATE without having to build exotic interface electronics into the test fixture.

Design signal interfaces so that inputs, outputs, and test point signals are at levels that are easily and accurately measurable (see Figure 4-1 for a preferred and nonpreferred example). Designing circuitry to

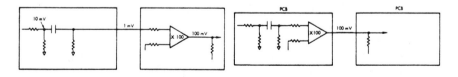

FIGURE 4-1. Analog signal interfaces. Measuring (or supplying) very small signals is a difficult task in analog circuits. Partitioning so that larger signals are presented at interfaces is one way to simplify the testing task.

provide reasonably large, reasonably robust analog signal interfaces will also minimize noise problems in the final designed unit.

High-Voltage Considerations

Designs that employ high voltages require special attention. Voltage dividers should be implemented between UUT circuit nodes and test points to limit the voltages (and currents) that are supplied to test points and to prevent damage to the UUT in the event that a test point is inadvertently connected incorrectly to an item of test equipment. Where accuracy is of a particular concern, precision voltage divider components, even if they cost a little more, should be designed in.

It is also good practice to design voltage divider networks so that there is one resistor connecting the node under test to the test point and two resistors of equal value, connected in parallel, connecting the test point to analog circuit ground for redundancy. Therefore, if one of the resistors connected between the test point and analog ground does open, the voltage on the test point will only double (instead of floating at the full potential of the node being monitored).

Provision should also be made for safely discharging such high-voltage circuits as pulse-forming networks. It would be nice to depend on every technician's regard for his or her own life in every instance, but the reality is that humans make errors. Especially in this day and age of litigation over product safety issues, every effort should be made to design systems that are as safe as possible.

Analog Metering

It is sometimes better to use an analog meter for certain system monitoring functions, especially where it is desirable to give users an approximate indication of system status. If an analog meter is selected for a design, the following guidelines should be considered:

- Banded meter faces (Figure 4-2) should be used throughout to preclude any unnecessary repairs because of noncritical functions.
- RF meters in drive-critical circuits should be placed as close as possible to the driver element for detection of feedline and voltage standing wave ratio (VSWR) problems.
- A bank of meter drive circuits should be used for allowing all metering to be center-lined after test verification at the system level (Figure 4-3).

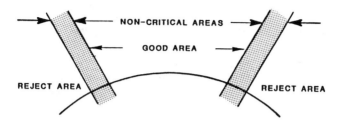

FIGURE 4-2.   Banded meter faces. Identifying noncritical areas via a banded meter face alerts users that the product is due for calibration at the next scheduled maintenance time.

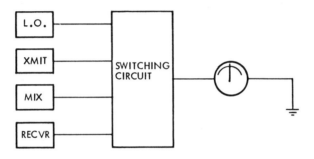

FIGURE 4-3.   Multiple-use meter circuit. Switching circuits should include the ability to center-line each subcircuit's correct operating point during calibration.

Analog Test Points

Just as test points in digital circuits provided very high leverage, thereby reducing troubleshooting times and costs by up to 50 percent per fault, and allowing for drastically reduced ambiguity group sizes in the event of a fault, so do analog test points. The same basic guidelines apply to the selection of analog test points, although one must often be more careful and rigorous when implementing analog test points.

To recap, if a measuring (test) instrument of any type was connected to an analog circuit node during design verification and debug, that node should be made a permanent test point that is accessible to both manufacturing test and service equipment and personnel. Next, use test points to separate analog and digital sections of the circuitry in order to provide for separate testing of those sections if the test strategy so dictates that approach. Then use the "cut it in half" method to further partition the analog circuitry and make fault isolation faster and more accurate.

If there are known critical nodes that must be monitored for calibration or fault isolation, they should also be made easily accessible test points.

In addition to the general guidelines for test points already mentioned, the following analog test point guidelines should be adhered to where they are applicable when designing new products:

- Do not allow test points to load down the signal or performance when connected to a measuring device (meter, etc.).
- For RF (radio frequency) and IF (intermediate frequency) signals, test points should be compatible with common impedances and connector types (50 or 75 ohms, BNC or SMA connectors).
- Evaluate the inclusion of test points for both test signal injection and signal observation.
- Never allow test points to represent hazardous voltages or high RF levels.
- Use BNC-type connectors for IF test points. This will make connection easier.
- Provide nearby ground points or turrets for the convenient connection of instrument and test probe return lines.
- Test points should provide a reasonable and useful facsimile of the signal being monitored and be designed for correct impedance matching.
- Test points can be important on packaged (throwaway) modules because they may provide the only means to isolate a fault to that module once it is installed in a circuit.
- On plug-in modules, test points are more accessible to ATE if they are built into the I/O connector.

An RF or IF path with a frequency conversion is very difficult to test under swept frequency conditions. The filter characteristic in the module shown in Figure 4-4 cannot be displayed on a network analyzer without a complex, external frequency converting test fixture. This problem can be corrected by installing a jumper or electronic switch between the amplifier and the filter and providing a visibility test point at the output of the filter (if its signal level is consistent with the previously mentioned guideline on signal interface levels). Otherwise just the partitioning jumper is sufficient, and the signal output can be measured at the subassembly's normal circuit output location.

Test points placed in metering circuits allow measurements to be made automatically (just like test points at the interface between digital logic circuits and indicators such as LEDs) and in a reduced amount of

Non-preferred Configuration

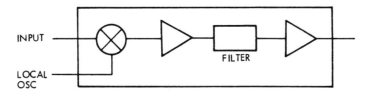

FIGURE 4-4. Unpreferred RF partitioning example. Mixing multiple frequencies in one assembly sometimes presents measurement problems. A jumper between the first amplifier and the filter will allow for easier filter alignment.

test time. In addition, such a test point capability requires a minimum number of tests to verify that the meter functions correctly (Figure 4-5).

If it is not possible to physically bring out as many discrete analog test points as are desired (or required) in a given design, it is almost always possible to utilize analog FET switches and/or analog multiplexers to bring multiple analog test point signals out to an analog testability bus connector. Alternatively, analog signals may be converted to digital data and made accessible through whatever digital testability interface has been included in a mixed analog/digital design.

Obeying fan-out and buffering guidelines is also important to insure that test points and UUT circuit outputs can drive the additional capacitances and lead lengths incurred with the use of extender cards and the connection of test equipment.

Particularly in analog and RF circuits, try to desensitize the I/O design (e.g., the partitioning) of individual assemblies in order to preclude testing mishaps. Designs should be immune to the transients, potential shorts, open circuits, and other conditions that can be encountered through testing. No test point should be excessively vulnerable.

Watch out also for circuit designs that promote new failures during system integration and field service troubleshooting operations. Figure

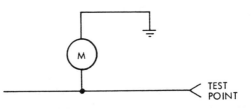

FIGURE 4-5. Test point on analog meter circuit. Electrical access to the circuitry driving an analog meter allows circuit performance to be verified without human intervention.

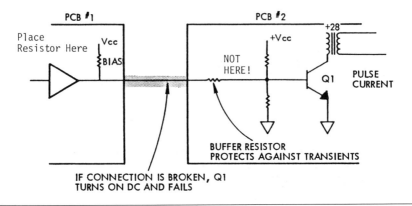

FIGURE 4-6.   Bias resistor placement. Bias resistors should be placed in such a way that removing one circuit board causes the other board's circuitry to turn off, not on (which could cause damage due to thermal runaway).

4-6 illustrates the right and wrong places for the location of a bias resistor in a low-duty-cycle pulse amplifier. If the bias resistor is located on PCB 2 and PCB 1 is removed with power still applied (do you turn the power off *every time?*), transistor Q1 will go into thermal runaway and fail. Thus the bias resistor should be placed on PCB 1.

A product is only as good as its user's ability to take advantage of its features. There are many true stories of truly feature-laden products that were absolute bombs in the marketplace because, no matter how technically elegant the products were, users could not use them! Testability features are far too often sacrificed so that extra functional features can be added to a product design—features that may actually delay time to market and make a product harder to use and harder to maintain. That trend must change if products are to be as successful as they can be and if their manufacturers are going to stay in business.

ANALOG CIRCUIT ELEMENTS

In this day of large scale integration of both analog and digital circuits, and indeed of mixed integrated analog and digital circuits such as signal processors and hybrid telecommunication chips, discrete analog circuitry may be thought not to exist. Such is not the case. There are still many products that make use of the old standby analog circuits, and these circuits also need to be testable. The next sections deal with guidelines for making sure that testability happens in even the most mundane (in some people's opinion) circuits.

Analog Discrete Components

Discrete circuitry consists of elements such as resistors, capacitors, and inductors combined with active elements such as diodes, transistors, and other active components. And, though integrated microcircuits are replacing most discrete components, many of those integrated components still require that discrete analog components be connected to their package pins in order to complete their functions.

The complex integrated devices tend to get a lot of attention from both designers and test people because of their newness and complexity. But sometimes it is the simple stuff, so common that people don't pay much attention to it, that causes the big problems in manufacturing, test, and service. It is therefore important to treat each component, whether a VLSI device or a lowly transistor, as an opportunity for lowering product costs and improving productivity in all phases of the electronics design and manufacturing business.

Analog Functional Modularity

Functional modularity, or partitioning, can now be more easily achieved by using integrated hybrid and monolithic circuits. Many of these integrated circuits are also available as MIL-STD approved parts. Even industrially rated parts are processed and inspected in such a way that MIL qualification is readily obtained. In the commercial arena, defect rates have declined to very few defective parts per million.

Many circuits that were typically previously designed with discrete circuits can now be purchased as complete packaged functions. Some typical examples of integrated devices are operational amplifiers (multiple-packaged), voltage-to-frequency (V/F) converters, DACs, ADCs, and phase-locked loops. Wherever possible, utilize monolithic or hybrid analog ICs in new designs instead of discrete analog circuitry. Doing so will reduce parts count, increase reliability, and reduce manufacturing, test, and service costs.

FREQUENCY CONSIDERATIONS

Analog circuits tend to be segmented into three major categories in accordance with the method of signal transmission.

1. Low-frequency circuits, where the characteristics of the interface with the test system can be regarded as essentially a lumped circuit load.

2. High-frequency circuits, where RF shielding and transmission lines must be considered (i.e., between 50 kHz and 100 MHz).
3. Microwave circuits, where distributed parameter analysis is required (i.e., above 100 MHz).

Each category presents special problems in modularization and placement of test points and will be treated separately.

Low-Frequency Analog Circuits

Low-frequency analog circuits are characterized by test methods which are influenced by loading, but which do not necessarily rely on impedance matching within the UUT or interface design. A single-point grounding scheme is generally used, and the majority of measurements are taken with the ground point as a reference. For these circuits, ATE compatibility is determined primarily by functional modularity and test point placement.

Low-Frequency Analog Circuit Modularity. Functional modularity in low-frequency analog circuits can significantly reduce the number of test points required for fault isolation in assemblies by reducing maintenance to direct replacement of the faulty subassembly. If a given circuit function is distributed over several subassemblies, intermediate test points must be provided at the subassembly inputs and outputs that are not accessible, directly or indirectly, at the operational and power connectors of the system.

Low-Frequency Analog Circuit Accessibility. A circuit output is accessible when there is fan-out to two or more circuit inputs which provide independent paths to an accessible point. In this case, if a single fault is present, the location of the fault can be deduced from external symptoms present on the accessible nodes.

Subassemblies should be supplied with sufficient test points to permit fault isolation to a subassembly or to a small group of nonrepairable components. To simplify the test setup, the subassembly operational connector may be increased in size by 25 to 50 percent to accommodate the necessary test points. In some instances, this might dictate use of multilayer printed circuit boards, whereupon the added cost must be considered.

Some other possible solutions to the test access problems in low-frequency analog circuits include separate test connectors and probe

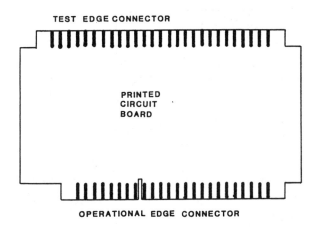

FIGURE 4-7. Test connector on alternative board edge. If there is not enough room on the regular card edge connector for the desired number of test points, an extra connector at the opposite end of the card for test purposes can often be implemented.

pads. A test connector mounted on a different edge (see Figure 4-7) may negate the need for multilayer printed wiring board construction.

Circuit pads provided for manual probing are not compatible with automated testing unless a simple interface device can be supplied with some means to contact these pads, thereby avoiding manual intervention during test execution. It is far more preferable to provide edge connector access, either with discrete test point connections or with a testability connector, especially in analog designs.

Low-Frequency Analog Circuit Test Interfacing. When low-frequency analog circuits are tested on ATE, it is the distributed capacitive load that is generally the most significant interface characteristic. Typically, this load is equivalent to a 200- to 1000-pF capacitor at the output of the UUT. This capacitance is usually distributed as follows:

1. External test cable: 50–100 pF
2. ATE interface device: 50–150 pF (each path)
3. Internal-ATE configuration cabling: 100–700 pF
    a. Unconnected to measurement instrument: 100–300 pF
    b. Connected to measurement instrument: 100–700 pF

Therefore, each and every test point cabled to the ATE may be loaded with roughly 250 pF, whereas the particular test point under

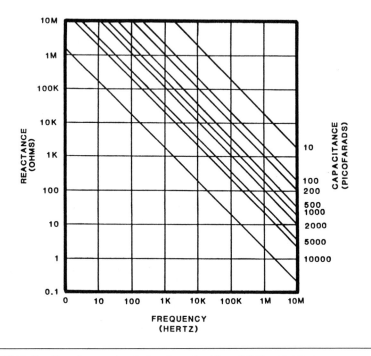

FIGURE 4-8.   Reactance chart. Analog circuits exhibit complex frequency-dependent impedances composed of resistance and capacitance. This chart allows for calculation of reactance at various frequencies.

observation may have perhaps twice that amount. Figures for bed-of-nails fixtures used with in-circuit and combinational testers are lower, but not by orders of magnitude. It is therefore doubly important that analog test points be properly isolated and buffered from the test interface.

The reactance chart in Figure 4-8 can be used to estimate the impedance represented by this capacitive load at various frequencies. A capacitance of 1000 pF and 10 kHz, for example, has a reactance of 16 kilohms. Thus, at 10 kHz, a 1-megohm AC voltmeter in an automatic test system can look like a 16-kilohm load at the UUT test point connector. Clearly, the effects of this load cannot be neglected in the circuit design and in the specification of nominal test values and tolerances.

Test points should be located at circuit nodes which are insensitive to the load imposed by the test cables. In a common emitter amplifier, for example, a small unbypassed emitter resistor provides a good test signal source which is relatively insensitive to loading and injected noise, but which is not short-circuit proof (see Figure 4-9).

Operational amplifiers, configured as voltage followers, also pro-

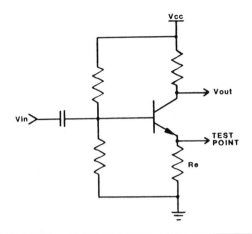

FIGURE 4-9.   Common emitter circuit test point placement. A test point placed at the emitter lead can provide valuable information about circuit function with little or no effect on circuit performance.

vide a good test signal source. There are many advantages to their use, as opposed to the emitter follower amplifier. Many operational amplifiers are internally short-circuit proof, and some are compensated for unity-gain operation. They can operate in a bipolar mode, and the output of a unity-gain buffer will be within a few millivolts of the input.

Often ATE can interface directly with UUT signals from operational amplifiers, but in cases where the signal range is too great, voltage division is required.

When operational amplifiers or emitter followers are used, they must be stable under the required loading conditions presented by the ATE or the circuit to be injected with a signal. An emitter follower can oscillate when connected to capacitive loads greater than 200 pF, and capacitive loading can add a phase shift to the feedback loop of an operational amplifier. This can cause peaking, ringing, and even oscillation at low amplifier gains.

To prevent the undesirable effects of capacitive loading, it is sometimes necessary to decouple the load line from the amplifier. This is done in an emitter follower by placing a small resistor in series with its base or the load. With operational amplifier circuits, the capacitive effect can be minimized by any of the following:

1.   Choosing an amplifier with low output impedance
2.   Adding a booster stage
3.   Using the phase compensation scheme in Figure 4-10

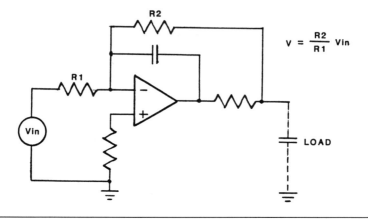

$$V = \frac{R2}{R1} \; Vin$$

FIGURE 4-10.   Phase compensation scheme. Capacitive loading caused by test equipment may require that analog circuit buffers include a phase compensation resistor-capacitor network.

When ICs are used, the overall parts cost is lower if the added circuits are included in the same package as the functional operational amplifier circuits. In such circuits, the major costs are in testing and fabrication. As long as the added chip surface area and power dissipation are small compared with the area and power required by the operational circuits, the additional costs are minimal. However, if the power dissipation or surface area is large, the IC yield may be reduced to the point where this approach is not feasible. With today's levels of integration and fabrication processes, however, a little added area is not usually a problem.

A decision must also be made on whether to include active buffers in a UUT or in the interface design. When added to the UUT, the weight and overall costs are increased but the test interface requirements are reduced. In addition, circuits within a UUT must undergo all of the severe environmental stresses required of the rest of the equipment, which would not be a requirement if the circuits were located in the interface design. The trade-off is one of reliability and cost-effectiveness. With today's technology in analog LSI and mixed analog/digital LSI and VLSI chips readily available, including semicustom circuits, providing test circuits on the UUT is normally easily achievable and provides additional production test capability.

When the equipment uses discrete rather than integrated circuitry, it is usually more cost-effective to place these active circuits in the interface design. Comprehensive test features can be built into the circuits within an interface adapter. These added circuits can, however, impact the usefulness of the interface design with other types of UUTs.

Adding Active Testability Circuits.   Active circuits added to a design have one function—to improve testability. Therefore, they must be added in a way that does not introduce new testability problems.

Clearly, when circuits are added in the form of additional discrete components, they may have an adverse effect on calculated reliability at the next level of assembly. The added circuits also constitute a functional part of the next lower assembly and, as such, must be tested along with all of the other circuits on that assembly.

When it is necessary to add additional circuitry to preliminary designs in order to allow sensitive output points to be tested, or to allow for injected signal generation into the UUT, the added circuits should be on the same subassembly as the components being tested. All circuits added for testability purposes should be made as simple as possible while still performing their intended function.

They should also be at least (and preferably much more) as reliable than the circuits that they are measuring and, if available, all unused gates, amplifiers, or other assembly circuits should be utilized for this purpose before adding new circuitry. Always investigate the possibility of adding additional circuitry to the design as an alternative to increasing the complexity and cost of the ATE system and interface necessary to perform a thorough test of the UUT. Finally, existing BIT/BITE circuits may often be utilized to help diagnose and perform various phases of automatic testing.

Discrete Versus Integrated Circuits.   The simplest discrete-component buffer amplifier is an emitter follower consisting of a transistor and an emitter resistor (two discrete components). Base bias for this circuit must be obtained by direct coupling to the circuit under test. Direct coupling, however, means that faults in the buffer circuit could be coupled into the circuit under test. Therefore, if an erroneous reading is observed at the UUT output and also at a test point, the suspect circuit and buffer circuit might both be replaced. Clearly, a simple two-component circuit is not very useful in reducing the ambiguity group size to four component parts. More complex buffer amplifiers, which can fail without affecting operational circuits, add new ambiguity groups with more than four parts.

The conclusion is that discrete component circuits are inherently difficult to fault isolate with functional ATE. IC assemblies are, therefore, much more compatible with fault isolation nonambiguity requirements.

Throwaway Maintenance Considerations.   One possible system or subsystem level maintenance strategy is not to repair failed subassemblies, but to simply discard them when they fail. The ability to use this strategy depends heavily on the cost of the failed assembly. With the increasing use of more complex and therefore more expensive devices and assemblies, numerous circuits are not candidates for throwaway maintenance. In this case, three alternatives are available:

1. Develop new IC designs.
2. Improve the cost-to-reliability ratio to a point where the subassembly qualifies as a throwaway.
3. Employ a support concept with automatic fault isolation on ATE only to the active element group and with off-line isolation to a component part using depot level equipment.

Combining Analog Test Points.   Most electronic circuit packaging schemes impose pin limitations which may preclude direct access to a sufficient number of test points for the required level of fault isolation. Most designers, when faced with pin limitations, reduce the number of test points to make the circuit fit the package. There is seldom any consideration given to combining test point signals. The resulting design is often incompatible with automatic test because there is insufficient access for fault isolation.

Figure 4-11 illustrates a method of achieving improved testability by combining, rather than eliminating, test points. In this example, a resistor summing network is used to combine two signals of opposite

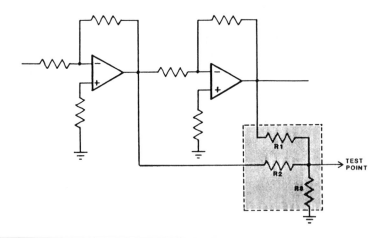

FIGURE 4-11.   Combining analog test points. Analog test points can sometimes be combined using a simple resistor network if there are not enough pins available at an edge connector or test connector for discrete test points.

polarity. Resistor ratios can be selected so that the polarity of the test signal indicates which amplifier is faulty.

The only requirement for combining test points is that at least one measurable characteristic of the combined signal be a function of each input signal. AC components of the output test signal are proportional to two different input signals. Many signals can be merged to one test point pin by using IC analog multiplexers.

HIGH-FREQUENCY ANALOG CIRCUITS

The high-frequency circuit category includes all circuits which generate or process signals with frequency components in excess of 50 kHz, yet below the UHF band (below 100 MHz). In this range, signal transmission via flexible coaxial cables is necessary for all but the shortest signal runs (e.g., more than a fraction of a wavelength).

Impedance-matched coaxial transmission is used to insure satisfactory reproduction of the input waveform at the receiving end of a transmission line. Line lengths in an automatic test system range from approximately 2 feet for a direct connection to a test instrument to 20 feet or longer for a switched connection via the interface configuration switch and internal system cabling.

Impedance-Matching Networks

Resistive attenuators are the most commonly used matching networks where wide bandwidths are required. The resistors in the network must have low series inductance and shunt capacitance with maximum permissible values determined by the desired frequency range. A typical impedance-matching network is shown in Figure 4-12. For those appli-

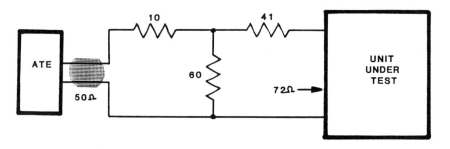

FIGURE 4-12.   Resistive impedance matching network. Depending on the frequency range of the signals being measured, a simple resistor matching network may be all that is required to match the impedance of the unit under test to the automatic test equipment.

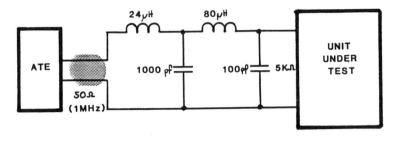

FIGURE 4-13. Reactive impedance matching network. In higher-frequency circuits, inductors and capacitors may have to be used in place of simple resistors in the design of impedance matching networks.

cations where a single frequency or narrow band of frequencies is involved, reactive LC matching networks can be used (Figure 4-13).

The methods and techniques for designing impedance-matching networks should be familiar to most circuit designers and can be found in standard engineering handbooks. Caution should be exercised because the standard handbook variety LC matching circuit is designed for maximum power transfer. This is desirable at the inputs and outputs of a UUT. Matching networks at the intermediate (fault isolation) test points, however, must be designed for minimum disturbance to the circuit under test. This means that the power withdrawn from the circuit must be minimized and that the matching circuit must not disturb the UUT by the open stub that is present when a test point is not switched to a response monitor instrument.

Test Point Isolation

In an ATE system, only one measurement at a time is taken. The response monitor inputs to the ATE are terminated in a characteristic impedance (Z_0) only while a measurement is being made. At all other times, the line from a UUT test point could be terminated at an open switch contact on a switching unit. The effects of these open stubs on the circuit under test cannot be neglected. Matching circuits within the UUT must isolate critical tuned circuits from the detuning effect of the reactive impedance of these stubs or must be compensated for in the interface design.

Two commonly used techniques for test point isolation are shown in Figure 4-14. The number of turns in the test point winding in the RF transformer is selected to minimize the effect of the ATE stub on the circuit under test while supplying a measurable signal amplitude that has a satisfactory signal-to-noise ratio.

When it comes to implementing BIT for high-frequency analog

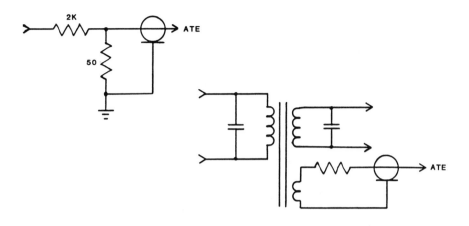

FIGURE 4-14. Test point isolation. To avoid loading a functional circuit and affecting its performance, resistor networks or transformer taps may be used to isolate a test point from the circuit under test.

circuits, it is often preferable for the BIT circuitry to deal with DC levels rather than RF signals. DC levels can be easily converted either to go/no go single bits (using a simple voltage comparator) or to actual values (using an ADC). An example circuit for performing the conversion of RF signals to DC levels is shown in Figure 4-15.

ATE Compatibility Design Guidelines

For maximum compatibility with ATE, the designer of RF circuits should adhere to the following guidelines:

- Input and output ports should be impedance-matched to a controlled impedance transmission line. In all cases where a different impedance level is used for operational signals, test points with design-specified impedance must be added.

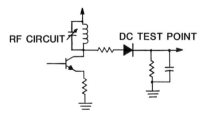

FIGURE 4-15. Converting RF signals to DC levels. Especially for built-in test purposes, converting RF signals to DC levels simplifies their monitoring.

- Test points should be designed for minimum disturbance to the circuit under test.
- The subassembly that houses the mixer circuit should also contain the local oscillator and at least one stage of IF amplification.
- Circuits should be testable with simple amplitude-modulated or frequency-modulated stimuli. Avoid designs which can only be tested with complex synchronized stimuli or which require random or pseudorandom audio-frequency modulation.
- Use prealigned frequency selective networks. Avoid the need for time-consuming manual alignment on the ATE station.
- Avoid continuously variable tuning controls which must be set by the ATE operator. Digital or contact-closure controls can be automatically controlled by ATE switching or digital stimulus generators. Digital synthesizers, for example, are more compatible than *LC* oscillators designed with variable capacitance or inductance tuning.
- Controls and indicators should be contained on an assembly separate from the one that contains the electronic circuits. This permits fully automatic test on ATE of the electronic part and manual test of switches and indicators with standard general-purpose instruments.
- Control switches which are an integral part of an electronic assembly should have a test position that permits automatic remote control of the electronic circuits by the ATE station.
- Provide means for opening AGC, AFC, and other feedback loops.
- Configure the mechanical breakdown of the equipment on a functional basis. No signal output of any subassembly should require processing by another subassembly for a return input to the original subassembly.
- Specify waveforms and frequencies at all interface points. The limiting range and tolerances should be accurately specified.
- Use standard connector types for all subassemblies with a uniform signal-to-pin number interconnection procedure: for example, pin 1 always ground, pin 2 always +VDC. This procedure permits interface adapter commonality with a minimum of interface switching.

Microwave Equipment Guidelines

In this text, the microwave spectral region is considered to be that portion of the electromagnetic spectrum where distributed parameters are of greater utility than lumped parameters. The frequency range will include all signals from approximately 100 MHz to 40 GHz. Current activity in microwave ICs has extended the use of lumped parameters to frequencies that are much higher than previously possible. Such devices and systems will continue to be considered in the microwave region and within the groups of UUTs in this section of the text.

ATE Compatibility Guidelines.   The microwave designer (like all other UUT designers) should

- Perform a test requirements analysis
- Evaluate the pertinent capabilities of ATE
- Design for best utilization of ATE capabilities
- Configure interface equipment that adapts the UUT to the ATE

The microwave designer, however, is confronted with the peculiarities of working in a distributed parameter environment. Desirable test points are frequently achievable only by relatively large structures. Test point accessibility is also a problem. Connections are not apt to have flexible leads but instead are more likely to be in the form of relatively rigid transmission lines that are sometimes quite large and usually precisely dimensional. Measurements are also a problem, since they are strongly dependent on the type and placement of sensors.

Whenever a sensor cannot be accommodated at the exact desired location in the UUT, care must be exercised so that the selected location does not significantly alter the parameter to be measured by the insertion loss and vector sum of all reflections in the required transmission line.

ADDITIONAL GENERAL ANALOG CIRCUIT GUIDELINES

In addition to the specific guidelines presented for specific categories of analog circuit designs, there are some general guidelines that apply, in principle, to all analog circuit designs. Only their circuit implementation is different (depending upon the frequency and voltage ranges employed in the design). These additional general guidelines include the following:

- Allow for the impedance loading of the ATE at the inputs and outputs of the UUT.
- Blend or combine output analog signals into a common test point if they are differential in nature.
- Describe circuits in terms of transfer functions by CAE/CAD workstations.

- Although a component may fail catastrophically, try to design the system to continue to function under degraded conditions.
- Where a failure may affect a large ambiguity group (amplifiers, feedback loops, AGC, etc.), several test points should be provided.
- A fault in a system should extend toward the output only and leave the input unaffected.
- The greatest voltage change from the normal generally occurs at the faulted component. Voltages should be charted for all nodal points.

TESTABILITY GUIDELINES FOR HYBRID CIRCUITS

Hybrid, or combination digital and analog, circuits present additional challenges in testing, and their testability can be most improved by partitioning the analog circuitry from the digital circuitry, either by physically partitioning them on separate assemblies or by partitioning them with appropriate test points and control and observation circuitry and logic.

Some additional specific guidelines for making hybrid circuits easier to test include the following:

- If glitches or spikes occur in voltage-to-digital conversions, and vice versa, try current-type conversions instead, because these tend to be smoother in operation and easier to test.
- R/R2 ladder networks are preferred because only two values are used, temperature coefficients are easily matched, ratios are easier to control than absolute resistance values, and component count is minimized.
- If there is a probability of a change in the signal during the conversion period, use a sample-and-hold technique to avoid digital miscounts.
- The greater the number of bits involved, the longer the conversion time will be. Use the smallest number of bits consistent with the desired result for speed and accuracy.
- Test points should be provided for reference voltage, sample and hold, phase-lock analog voltage, and voltage-controlled oscillator (VCO) outputs.

Figure 4-16 illustrates placing test points between the digital and analog circuitry in a hybrid circuit design.

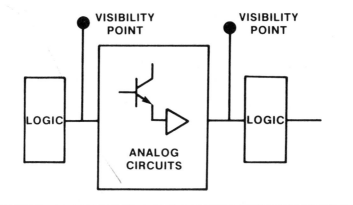

FIGURE 4-16.   Test points in hybrid circuits. Test points in hybrid (i.e., mixed analog and digital) circuits should be placed at the interfaces between digital and analog circuit segments.

Electromechanical Interfaces

It is also recommended that test points be placed at the inputs and outputs of circuits that interface with electromechanical devices, such as relays and loudspeakers. Figure 4-17 illustrates the placement of test points in these instances.

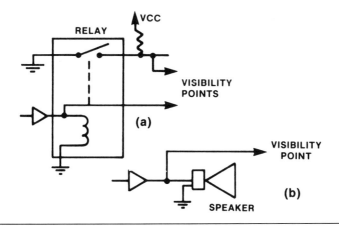

FIGURE 4-17.   Test points for electromechanical devices. The operation of the circuitry driving electromechanical devices can be more easily verified if some electrical visibility to it is provided.

Electro-optical Interfaces

Interfaces between digital circuitry and electro-optical circuitry, or between strictly digital circuitry and hybrid circuits designed to interface digital circuitry to optical displays (such as CRTs), are also likely candidates for test points. Figure 4-18 illustrates the placement of test points at the outputs of LED drive logic (a) and at the inputs to circuits such as video encoders (b).

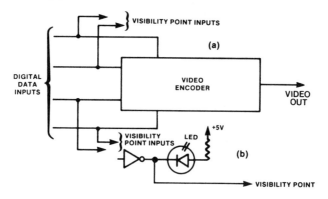

FIGURE 4-18.   Test points for electro-optical circuits. The operation of the circuitry driving electro-optical devices can be more easily verified without human intervention when some electrical visibility is designed in.

5

LSI/VLSI Board
Level Guidelines

Semiconductor manufacturers are now placing on the chip what used to be contained in an entire system. Industry statistics indicate that the number of devices per PCB is not decreasing. Rather, the number of functions per PCB is increasing in order to meet marketing and application demands. The result is that the average 100-IC PCB is far more complex. As the complexity of assemblies increases, the cost to test them increases exponentially (see Figure 5-1).

While the design engineering groups may have the latest development systems at their disposal, the test engineering groups are usually equipped with systems that are designed with components and technology that is at least one step behind the technology of the boards they are to test. And the ATE may or may not be properly linked to the design engineering CAE workstation.

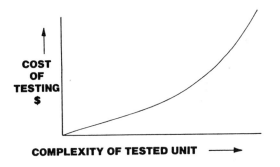

FIGURE 5-1.   Cost of testing versus complexity. While the complexity of a UUT increases linearly, the cost of testing that unit can rise much more rapidly.

It is important to remember that, when dealing with LSI and VLSI logic, the SSI and MSI guidelines do not go away. In fact, the guidelines for SSI and MSI circuits can be applied to the internal structure of LSI/VLSI devices at the chip level. The testability guidelines build on each other, and the fundamental concepts of synchronization, partitioning, initialization, control, and visibility must not be ignored for glue logic any more than for LSI/VLSI devices.

LSI/VLSI BOARD ADVANTAGES AND DISADVANTAGES

LSI/VLSI-based boards have advantages and disadvantages when it comes to implementing testability and dealing with their increased complexity. This section looks briefly at the advantages and disadvantages, as well as at failure mechanisms.

Advantages of LSI/VLSI-Based PCBs

Because LSI/VLSI devices have so much internal capability, they can be designed into a PCB in such a way that the PCB is inherently partitioned. It is important in the design for testability to take advantage of this feature. Figure 5-2 illustrates the typical structure of an LSI/VLSI-based board and the interfaces needed to external test equipment (or built-in test circuitry elsewhere on the board or in the system).

Once the LSI/VLSI-based assembly has been structured for partitioning, an effective testing strategy can be employed. One such testing scenario might be as follows:

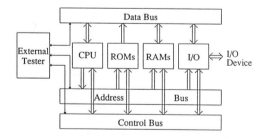

FIGURE 5-2.   LSI/VLSI-based board structure. A typical LSI/VLSI-based board contains a processor, ROM, RAM, and I/O devices. Designers should take into account the external tester by providing control and visibility to on-board busses.

1. Verify that the address, data, and control busses, with all components tristated, are free from any stuck-at-1 or stuck-at-0 faults.
2. Allow the ATE to communicate, via the bus, with the microprocessor and run a few basic OP codes.
3. The ATE should then check the read-only memory (ROM). A checksum comparison is a good test of this section.
4. If a self-test program is included in ROM on the PCB, the microprocessor can then be allowed to run the self-test and check the random access memory (RAM) and any remaining devices on the bus while the ATE is monitoring the test

Another advantage of LSI/VLSI-based PCBs is that test points on the bus are common to many devices and provide access to all of them, hopefully one or a few at a time, via a shared arrangement. This feature greatly reduces the fixturing and interface requirements to the ATE and better defines the selection process for test points.

LSI/VLSI-based assemblies containing both a microprocessor and on-board memory (typically ROM) have self-testing capabilities. Self-tests are (typically) small programs, usually less than 2,048 bytes, provided on the PCB as a means of verifying that the PCB is performing all of its functions. The on-board test program can be used in conjunction with the ATE to generate multimillion pattern test programs with little problem.

Disadvantages of LSI/VLSI-Based PCBs

By their very nature, LSI/VLSI devices are inherently sequential—the entire PCB of yesterday has been implemented on a chip. Long counter chains and deep sequential networks are an integral part of many LSI/VLSI devices. Since the engineer designing the assembly into which commercially available LSI/VLSI devices are being designed typically has no control over the inner workings of the devices—unless they are custom devices—testability becomes a problem.

Another disadvantage stems from the fact that most LSI/VLSI devices are designed to work within a bus-oriented PCB architecture. Thus most operate with three input/output (I/O) levels: logic low, logic high, and high impedance (tristate). Many of these devices also operate in three modes: drive, receive, and "off" (tristate).

Where once it was always possible to ascertain the direction of signal flow based on the logic design, LSI/VLSI devices complicate things by changing modes, over time, under control of software. One test axiom in the SSI and MSI world was: replace the driver IC first (most likely failure) and then the driven IC. LSI/VLSI PCBs have a new problem—sometimes a chip is a driver and at other times it is a "driven" (e.g., a receiver). These new features greatly complicate the fault isolation process and are a major cause of increasing test costs.

PCBs with LSI/VLSI devices tend to be large feedback networks via hardware, where events are caused to happen, say, by a microprocessor, and later evaluated by the same microprocessor, or via software, where the result of an event is determined by a previous event, often internal to the chip. Again, since the design engineer has no access to the internal workings of the chip, he or she has a testability problem.

Finally, the very bus structure that helps with partitioning creates new fault isolation problems due to the wired-OR nature of the busses. With multiple I/O devices connected to each bus line, it is difficult to isolate a failure on a bus line to an individual device. This is the reason for the guideline that follows for providing ATE (or BIT circuitry) with complete control over all device enable lines. With proper control, it is possible to diagnose faults beyond the node and down to the exact faulty device causing the failure on a board.

LSI/VLSI Device Failure Mechanisms

Another thing that the advent of LSI/VLSI has provided is a new set of failure mechanisms. While the standard stuck-at-1/stuck-at-0 (i.e., shorted to either the V_{cc} or ground potentials) failure modes common to SSI and MSI devices still occur, a new set of failure modes has surfaced. These new failure modes are called *soft failures* and include pattern sensitivity, timing sensitivity, noise sensitivity, and intermittent failures. This new set of possible faults has led to such new testing problems as the need for dynamic functional testing for operating the devices at, or close to, their rated operating speeds, and the need for very long (multithousand to multimillion step) test patterns so that all possible fault conditions may be detected.

As device complexity continues to climb, it can become literally impossible to achieve high enough fault coverage at the device level to ensure that all functions within a device are operating correctly.

There are, in fact, several instances where microprocessors and other devices, even when designed with some testability features, were found to contain latent design and/or fabrication defects that caused problems for users that were not detected by the device manufacturers' final functional tests on the devices. So while both merchant and in-house device designers wrestle with making the devices themselves more testable, it is important that the device user—the LSI/VLSI printed circuit board designer—makes sure that the device is as testable as inherently possible *when the devices are installed on boards.*

PARTITIONING OF LSI/VLSI-BASED BOARDS

It is important to allow isolation of board subsections for partitioned testing. The most difficult job in partitioning a microprocessor board for fault isolation is sectioning the bus. An example of a multiprocessor-based board with minimal partitioning is shown in Figure 5-3. This board is part of an electronic typewriter and contains four processors, four free-running clocks (oscillators), mixed analog and digital circuitry, and minimum I/O connections. It is an extremely untestable configuration.

If the busses are not functional, nothing will operate. And there is no way, given the current design, to remove any of the processors from the bus in order to determine where a fault may lie. The ROM data in each of the 8048 processors cannot be disabled and replaced by data supplied by an ATE system. The four free-running clocks make it difficult, if not impossible, for an ATE system to synchronize to the board for testing purposes—in short, not a pretty sight for the test engineer tasked with developing a test program for the board.

To make the board at least minimally testable, certain design

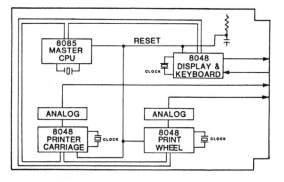

FIGURE 5-3.   Untestable multiprocessor board. A board with multiple clocks and one master reset has no control and no partitioning and is therefore very difficult to test.

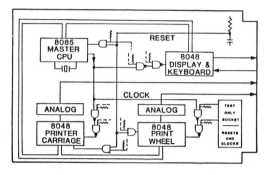

FIGURE 5-4.   Minimally testable multiprocessor board. A few extra gates and resistors added to a complex design can make it much easier to test, even if a special test connector or socket is required to access them.

changes must be made as shown in Figure 5-4. The first of these changes is to break up the single (currently uncontrollable) power-up RESET line and allow a tester to reset each of the processors on the board individually. This change immediately divides the problem by a factor of 4 and allows each section of the board to be tested individually prior to letting the board's sections work together in their ultimate functional configuration.

Then the multiple free-running clocks should be replaced with a single controllable master clock for the main processor that can also drive the subsidiary processor clocks. And the clock inputs to each of the 8048s should be gated (since the single step lines for those chips are not available for test access) so that those processors can be tested at ATE, rather than UUT, speed.

Finally, the address, control, and data busses have been made available at the edge connector of the board. There was insufficient room at the edge connector for the additional reset and clock inputs, so a test-only socket was added to the board. The ATE system plugs into that socket during testing. While it would be better to have brought those lines to an edge connector interface (either directly or through some testability circuitry) to eliminate additional test handling time for connecting the extra cable and to implement some of the many other guidelines that follow in this chapter, the changes did make it possible to test the board, and the economic results were impressive.

Table 5-1 summarizes the overall product costs by major category for both the untestable and minimally testable versions of this design. As the table shows, the return on investment through incorporating the

TABLE 5-1. Cost Comparison with Testability

Cost Element	Without Testability ($)	With Testability ($)
Test Equipment	500,000	310,000
Test Program	31,550	18,400
Test Fixture	20,500	14,450
Diagnostic Cost/Board	22.41	12.14
Parts Cost/Board	101.34	103.68
Cost to Make Design Changes: $100.00		
Cost Savings at Unit Number 1: $209,200		
Cost Savings over 5-Year Run of 50,000 Boards: $605,700		
Reliability Decrease: 0.00012%		

testability features was phenomenal. Over the five-year life of the product, the total savings were over 600 thousand. The nonrecurring cost savings at the prototype stage (i.e., unit number 1) were $209,200—all on an investment of $100! Parts costs over the life of the product were increased about 2 percent (from $101.34 to $103.68), but the large savings in test costs, particularly in the area of diagnostic cost per board, more than paid for the small increase in parts costs.

What about reliability? The calculated (and actual) reliability decrease due to the changes was so small as to be almost infinitesimal. *The return on investment is in the millions of percent.* This example illustrates the kind of leverage that can result from incorporating testability into complex designs.

In general, if the microprocessor and other chips that normally drive the bus can be put in a high-impedance mode (i.e., tristated) and if tester access to the bus is provided, the bus can be tested as an entity. Board partitioning makes a complex testing task realizable by allowing the test program to be sectioned or structured.

A well-structured testing philosophy is always contingent on being able to partition the LSI/VLSI-based assemblies under test. The advantages of tristate conditions and the inherent partitioning of LSI/VLSI-based assemblies are completely lost if the PCB is designed with these lines tied hard to a power or ground bus. Tristate control lines should be made alternately controllable by the ATE wherever possible.

Partitioning is an effective technique for breaking feedback loops and provides a means to isolate faults within those loops. Single-board computer architectures typically have large feedback loops formed

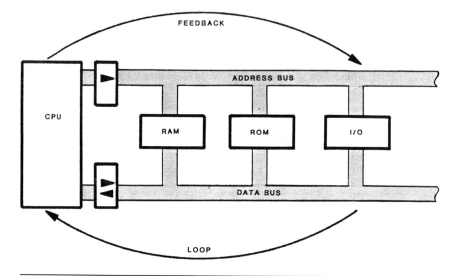

FIGURE 5-5.   Processor board feedback loop. Because of the connections between the address and data busses through most of the devices on a processor board, a large feedback loop is often created.

through the address and data busses. Any error would be easy to detect because all data streams within the loop would appear to have erroneous data. However, the fault isolation would be very difficult. Refer to Figure 5-5.

Figure 5-6 shows a method for breaking the classic single-board computer feedback loop. This can be accomplished at either the address or data bus (preferably at both). In this example, an AND gate is added to logically "and" the BUS REQUEST line with a control signal from the

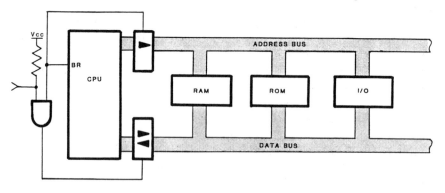

FIGURE 5-6.   Breaking processor board feedback loops. External control of unidirectional and bidirectional latches and buffers through the implementation of an extra gate allows processor board feedback loops to be conveniently broken.

ATE. By pulling the control line to a logic 0, the tester may tristate the bidirectional buffer and break the feedback loop. This will allow quicker and easier fault isolation.

CONTROLLABILITY OF LSI/VLSI BOARDS

Controllability is the ability to externally (typically via an ATE) alter the internal status of a UUT. Control is imperative if the board is to be functionally testable. Control is especially needed over processor lines such as those typically called READY, RESET, HOLD, TRAP, and NMI (nonmaskable interrupt).

A small section of a schematic for a microprocessor-based PCB is shown in Figure 5-7. In this example, program control is passed to the ROM whenever a reset or interrupt occurs. The normal operating program in this ROM then services the interrupt request.

During the testing process, it is desirable to transfer control from the on-board ROM to the ATE. This allows the ATE to have maximum control and lets the test programmer execute any needed special testing code with out losing control of the PCB.

An additional concern in this circuit is the requirement for the RESET line to be controlled directly by the ATE. In this example, the only way to initiate a reset is to cycle the power or to indirectly control it, if possible, by using the SYSTEM RESET line. Control is needed to aid in the initialization process and to allow the chip select decode circuitry to be verified.

The minimum recommended improvements for testability are shown in Figure 5-8. A three-input AND gate is used in place of the two-input gate that handles the power fail detector and the external system reset input. The third input is pulled high through a resistor, and a control point is made available to the ATE. This will allow the ATE to

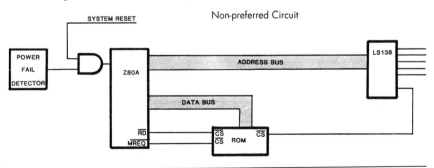

FIGURE 5-7.   Processor board with interrupts. External asynchronous interrupts can interfere with the testing process, especially if the on-board ROM cannot be disabled.

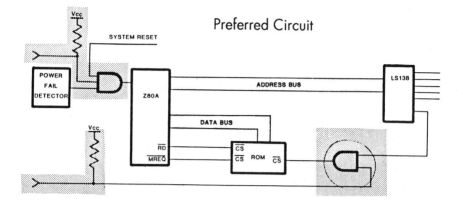

FIGURE 5-8.   Testable processor board with interrupts. A three-input gate instead of a two-input gate provides a means to disable external interrupts, and an added gate disables on-board ROM so that the ATE can emulate it for its own purposes.

initiate a central processing unit (CPU) reset with a single pulse. The now unused two-input AND gate can then be used to inhibit the INTER-RUPT signal, thus allowing the ATE to electrically isolate the on-board ROM and emulate its function through the data bus, which can now been freed.

Control is of vital testing importance and does not require much extra circuitry to achieve. As illustrated in the previous example, an extra three-input gate and two resistors added to a very complex board can make the testing task far easier, faster, and cheaper with virtually no impact on circuit board size, parts cost, or reliability.

Another example where proper controllability has not been provided is shown in Figure 5-9. This is a section of a microprocessor-based

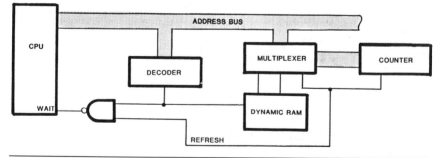

FIGURE 5-9.   Processor board with dynamic RAM. Dynamic RAM requires periodic refresh cycles which can interfere with the testing process.

PCB containing dynamic RAM and its associated refresh circuitry. As can be seen, no methods are in place for controlling the WAIT line. This line is important from a testing standpoint for two reasons. First, allowing control over the WAIT line permits the use of an I/O rate test system for testing this assembly. The alternative would be to use a very expensive high-speed dynamic test system which may not be available. Second, should a failure occur in the control or refresh circuitry, random wait states could be generated, or, worse, the entire system could be locked up, thus preventing efficient fault isolation.

The solution to this testing problem is to use a three-input NAND gate, as shown in Figure 5-10, in place of the two-input NAND.

The third input is tied to V_{cc} through a resistor, and control over that input is made available to the ATE. Now, under program control, the ATE can disable the effect of the refresh and control circuitry and prevent any failures in this section from locking up the system and hindering the testing of the other sections of the board.

As with SSI and MSI assemblies, edge connector access to the control points is recommended as a first alternative. As a minimum consideration, RESET and HOLD lines should be tied to V_{cc} or ground through an appropriate pull-up or pull-down circuit. Allow for IC clip or bed-of-nails fixture access when it is not possible to bring control lines to unused edge connector pins. If many control lines are needed on a given board, consideration should be given to providing testability circuitry that can drive multiple control lines via a few testability interface lines that can either be connected to a few edge connector pins or brought to a test-only socket.

Preferred Circuit

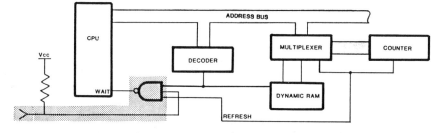

FIGURE 5-10.   Testable processor board with dynamic RAM. An extra input to a gate, controllable from an outside source, can be used to disable the periodic refresh request for testing purposes.

Additional control points increase the testability of an assembly in three ways. First, increased control can greatly reduce the amount of time and effort in the generation of test programs. By having external control of a circuit, fewer patterns will be needed to set up proper testing conditions and to propagate faults to a circuit node monitored by the ATE. Second, fewer test patterns due to increased control allow test programs to be executed in a shorter amount of time. Third, increased control reduces the number of possible fault types during any given testing operation and allows for faster fault isolation. As mentioned, the majority of time expended in testing most LSI/VLSI-based PCBs, with typical failure rates and fault distributions, is during the fault isolation process. A significant amount of that time is spent tracing the wrong fault because of multiple faults per PCB or because of faults masked due to lack of control which prevent an accurate diagnosis.

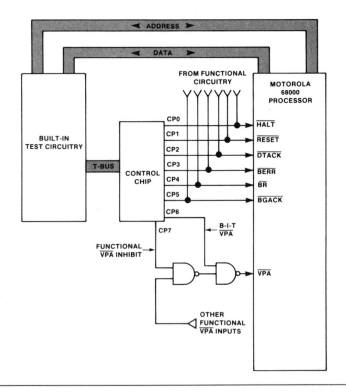

FIGURE 5-11.   Processor control with testability circuits. Control is often desirable for critical processor signals. Testability circuits can be added to a board to provide that control directly from an automatic tester (or built-in test resource).

Processor Board Controllability Using Testability Circuits

Recommended control points for many different types of processors and peripheral (or support) circuits may be found in the appropriate section of Chapter 6. The following example (see Figure 5-11) shows the implementation of a testability circuit to provide control of on-board processor lines via a testability bus connection for a board designed with a 68000.

Via a control type testability IC, the built-in test circuitry can control the operation of the processor at any time. When the ATE is not connected to the testability bus interface, the ENABLE line of the testability control circuit is held in the logic 1 state, which tristates all of the outputs of the testability chip. The control circuit can be a shift register, a multiplexer, or a combination of both.

As shown in the drawing, in many cases control point outputs from the testability chip can simply be connected in parallel with the normal functional circuitry. In a few cases, however, an extra gate or two may be required. The example shows the case where the valid peripheral address (VPA*) line requires two NAND gates and two control point outputs. In this application, the ATE system needs to be able (a) to inhibit any functional circuitry responses to the processor input and (b) to assert to the processor that it is itself a valid peripheral.

VISIBILITY ON LSI/VLSI-BASED BOARDS

Visibility is the ability to externally monitor the internal operation of a unit under test. With SSI and MSI PCBs the need for many test and control points exists (the rule of thumb being one test or control point per integrated circuit); with LSI/VLSI-based PCBs the number of these points is reduced as long as access to control, address, and data busses is available.

Visibility to certain lines on LSI/VLSI-based PCBs is now more standard and more crucial, especially for bus lines and status indicator lines. Address and data busses should always be visible to the test equipment or BIT resource.

The best access for visibility is again achieved through the edge connector or a test-only socket, since the speeds at which LSI/VLSI-based PCBs operate may preclude the use of IC clips or bed-of-nails fixtures. Access to keyboards and displays is especially important to eliminate human interaction and to reduce fixturing and interface problems. Even with LSI/VLSI-based PCBs, all of the previous guidelines for visibility apply.

Increased visibility increases the testability of an assembly in three ways. First, increased visibility can greatly reduce the amount of time and effort in the generation of test programs. By having internal visibility into a circuit, fewer patterns will be needed to propagate faults to a circuit node monitored by the ATE. Second, fewer test patterns allows test programs to be executed in a shorter amount of time. Third, increased visibility reduces the required number of operator probes during the fault isolation process. The majority of the time expended in testing most LSI/VLSI-based PCBs, with typical failure rates and fault distributions, is during the fault isolation process. The probing sequence is typically software guided by the ATE, which prompts an operator to probe the required points on failing assemblies. Due to the high level of human intervention, this is a potentially long process and quite prone to error. In many cases, the addition of one test point can cut the number of operator-probing operations required by a factor of 2.

The general guidelines for visibility points for LSI/VLSI-based boards are the same as those for any other circuit—critical nodes should be monitored first and then the "cut it in half " method should be used to choose visibility points.

Processor Board Visibility Using Testability Circuits

Just as it was possible to implement multiple control points on a board using only a few edge connector pins for a testability interface, it is also possible to achieve the same multiplication of edge connector pin leverage by using visibility-type testability circuits—again either shift registers, multiplexers, or combination circuits. The same 68000 used in the previous example is used in Figure 5-12 to illustrate the concept of monitoring critical processor output and status lines with dedicated testability circuits.

In this example, the seven most critical 68000 processor visibility points are connected, in addition to their normal functional circuitry connections, to the inputs of a visibility circuit. Then the ATE can look at any of these critical visibility points, either by latching data into the visibility circuit and clocking it out a single serial line or by addressing any individual node and monitoring it in real time. Thus, gaining visibility to virtually all of the critical points on the processor, as well as to any other visibility points on the board, can be accomplished using as many visibility point inputs as needed on the testability chip while still requiring the minimum number of physical I/O connections for testability at the board edge connector or test connector interface.

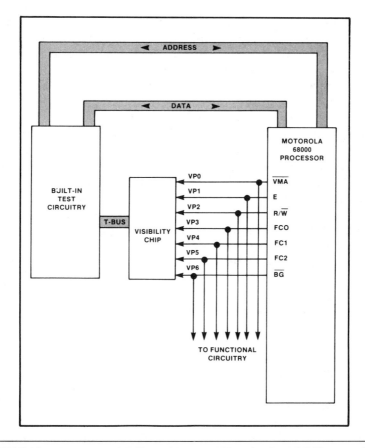

FIGURE 5-12.   Processor visibility with testability circuits. Visibility to critical processor output pins is often a testing necessity. Visibility can be accomplished by adding a monitoring circuit to a board and accessing it via a testability bus.

INITIALIZATION

Initialization is the process of getting memory elements on a logic circuit board into a known state. Digital testing and fault isolation and testing cannot begin unless the UUT is first initialized. Like SSI and MSI, LSI/VLSI also requires initialization. However, now there are internal memory elements which cannot be initialized directly via hardware. Remember that any fault which precludes initialization of the UUT cannot be diagnosed with the ATE.

Thus an important testability element to be considered that has been caused by LSI/VLSI technology is the requirement for software initialization. Software initialization requirements come about due to two factors. One is that LSI/VLSI-based PCBs are inherently sequential, and the other is that access to the logic elements internal to an LSI/VLSI device which require initialization cannot normally be easily achieved via hardware means.

The problem is compounded when an interrupt is serviced prior to software initialization. This interrupt may be generated by an uncontrollable programmable ROM (PROM) routine, with the result being that unknown states are propagated throughout the PCB. This precludes synchronization by the ATE. Adequate visibility and control points will allow a test program to perform software initialization.

It is important, therefore, to select devices for new designs that have reset (or equivalent initialization) lines and to make sure that those lines are controllable by either ATE or built-in test circuitry once the components are installed on the board. It is usually relatively easy to determine the required initialization lines by examining the pin descriptions listed on the device data sheet for the individual pins. Alternatively, the block diagram for the device may be consulted to find any lines that directly control the state of any on-chip latches or other memory elements.

SYNCHRONIZATION

The UUT and the ATE must be in synchronization with each other in order for testing and fault isolation to take place. This has recently become a major concern of engineers charged with programming ATE, primarily because many of the microprocessors are operating at frequencies higher than most ATE can handle.

The more sophisticated processors include internal clock generators and do not lend themselves to external control for testing when configured as listed in the manufacturers' data books. Most dynamic functional testers only operate at 10- to 20-MHz clock speeds, although a few ATE manufacturers boast of speeds from 30 to 40 MHz. However, even with the 40-MHz clock speed tester, the fastest that data can accurately be collected is 20 MHz.

Certain types of faults are best diagnosed at specific speeds: manufacturing faults at slow speed, pervasive timing faults at controlled dynamic (single-stepped) speed, and subtle timing or design faults at free-running speeds.

With access to the READY and HOLD lines, the speed of the microprocessor can be controlled by the tester. Single-stepping or microprocessor-isolated testing is also allowed, and with control of on-board

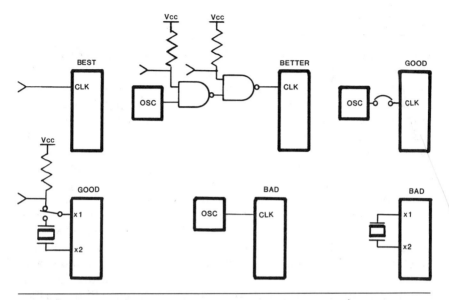

FIGURE 5-13.   LSI/VLSI board clock control examples. There are many ways to provide for disabling an on-board or on-chip clock and substituting an external clock for testing purposes.

clocks or access to on-board sync lines the use of high-speed ATE is possible.

As shown in Figure 5-13, clock lines should either come from the PCB edge connector or be enabled via control points activating logic inserted between the clock and its eventual destination. They can also be designed to be disabled and overdriven. In the case of on-chip clocks, buffer circuitry should be provided on the PCB to provide the ATE with a synchronization signal. If left unbuffered, connection of the ATE to the unit under test may disable the clock circuit or alter its operating parameters.

LSI/VLSI-based systems often require more than one on-board clock. These clocks need not synchronize with one another from a functional standpoint. Multiple clocks, however, can cause tremendous testing problems from all three testability standpoints: program generation, test execution, and especially fault isolation. If multiple clocks are required in a particular design, it is recommended that all lower-frequency clocks be derived from one master clock. Do not use multiple free-running clocks if at all possible, and remember to allow for master clock input/enable/disable as well as resets for the divide-by-N counters. In any case, it is crucial that there be synchronization between the unit under test and the ATE and that synchronization be both predictable and repeatable.

SELF-TESTS

Mention was made of multimillion pattern requirements for thorough testing of a complex LSI/VLSI-based PCB. To generate so many patterns for an ATE program is a formidable task. But the on-board PROM (1K or 2K) can take advantage of the microprocessor to help generate multimillion pattern test programs with little problem. The key from a testability standpoint is to make sure that the test engineer can take advantage of the self-tests.

Self-tests are generally designed to perform go/no go testing of the assembly in which they are resident. As more PCBs become equivalent to complete systems, they will benefit from (or require) self-tests.

In addition to providing system go/no go status, properly designed self-tests can reduce the test programming effort considerably. A well-designed 2K self-test in ROM reduces the test engineer's job to controlling the flow of data rather than generating a multitude of unique data.

The general guidelines for self-tests include structuring them for partitioning and writing "standard" routines. When many PCBs use like devices configured in ways substantially similar to the "kernel" of the PCB, which may have a totally different function or I/O structure, use self-test routines that are transportable from PCB to PCB for the devices used. This reduces the proliferation of self-test programs and simplifies the design engineering job. Many self-test programs need not be reinvented for each new PCB.

DEVICE STANDARDIZATION

Testability is greatly enhanced when multiprocessor PCBs use similar, or even better, the same devices. The test program generation effort is considerably reduced by the ability to use the same, or similar, program modules for multiple devices.

LSI/VLSI devices take a long time to fully characterize. Avoid the temptation to use the "latest and greatest" just for the sake of doing so. It may be overkill for a given application, it may not work, it may not be available long-term, or it may not be documented properly.

Specifications for SSI and MSI devices were easy—fan in, fan out, propagation delay, and so on. Specifications for LSI/VLSI devices generally follow similar formats, but the devices are orders of magnitude more complex. An idiosyncrasy that means nothing as far as the product performance from a design standpoint can cause nightmares in board test. Standardizing on a specific device, or family of devices, lets one understand the specifications far better and in far more detail.

Finally, when implementing LSI/VLSI devices on boards, particularly new state-of-the-art VLSI devices, do not use unsupported (from manufacturers' data sheet standpoint) instructions that may have been reported in the trade press or passed on to you by a colleague. Unsupported instructions have an alarming tendency to disappear as device manufacturers change masks and processes in their continual efforts to improve process and final device yields. The really great trick played by using an unsupported instruction could literally backfire and cause untold problems should that unsupported instruction cease to operate as before.

SUMMARY OF LSI/VLSI BOARD GUIDELINES

In summary, an LSI/VLSI-based board must be designed for partitioning, control, and visibility, with special attention to control of tristate lines, OUTPUT ENABLE lines, CHIP SELECT lines, and address, data, and control busses, along with clock control, initialization, synchronization, and prevention of unwanted interrupts during testing.

Control of HOLD, WAIT, SINGLE-STEP, and EXTERNAL ACCESS (or their equivalent) lines should be implemented either directly to an edge connector or via a test-only socket so that external ATE can be the dominating element during board testing and troubleshooting.

Clocks should be carefully controlled, and it should be possible for the test resource to inhibit on-board (and on-chip, where appropriate) clocks so that tester clocks can be substituted. Clock control is particularly important if the useful life of existing test equipment is to be prolonged and if tests are going to achieve adequate fault coverage.

Every attempt should be made to use testability and test program compatibility analysis tools to insure that design verification test patterns will run on ATE without extensive modification. Full fault simulations should also be run by the designer of an LSI/VLSI-based board to insure that both functionality and adequate test coverage have been successfully implemented. To do less is negligence.

Figure 5-14 is a block diagram of a typical LSI/VLSI-based board design where testability considerations have been pretty much ignored. In contrast, Figure 5-15 illustrates the testability changes that should be made to the generic LSI/VLSI-based PCB design. The design illustrated makes use of most of the technologies in common use today (and forecast for common use in the 1990s).

As will be apparent from the drawings, even the most complex board designs can be rendered very testable with only minimal addi-

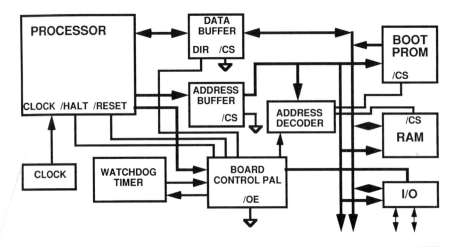

FIGURE 5-14.   Typical untestable LSI/VLSI board summary. The figure shows a typical processor-based LSI/VLSI-based board that was designed with no attention to testability. It cannot be partitioned, and there is minimal visibility and control.

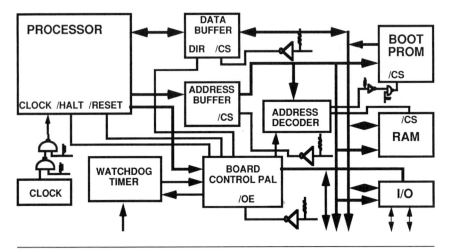

FIGURE 5-15.   Testable LSI/VLSI board summary. The addition of five gates and five resistors, along with some physical access to the busses and control lines on the board, provides for partitioning, control, and visibility.

tional circuitry as long as the proper partitioning and control and visibility access are implemented. Each design should be checked before it is committed to production to make sure that the required testability features are implemented.

6

Merchant Devices on Boards

This chapter covers the control and visibility points required for testable board design using many different microprocessors and several of the most frequently used peripheral chips. After becoming familiar with these guidelines, each designer should be able to extrapolate them and choose specific guidelines for any new device or subassembly.

Each device is briefly described, and some of the pinouts or block diagrams are also illustrated where appropriate. Each device is then examined with respect to clocking and synchronization to the ATE system, initialization (hardware and software, where applicable), status and data evaluation by the ATE, and any special considerations for the use of the device.

Since a microprocessor board is, in many cases, a complete product, it is usually not too difficult to write a small routine (using assembly language) to self-test a PCB. The speed of most ATE, however, restricts the ATE system's ability to effectively handshake (i.e., read and write synchronously) with a microprocessor board running at rated speed. It is important, therefore, that timing relationships be fully understood and taken into account, especially from a diagnostic standpoint.

Another consideration is to allow external ATE to emulate any on-board or on-chip ROM. If the test resource can accomplish this, it can use the power of the UUT itself to accomplish some of the testing. RAM, for example, can be tested in one of two ways: (1) by applying millions of test patterns over the address and data busses from the tester; or (2) by letting the tester emulate ROM and telling the processor on the UUT, using a dozen or so instructions, to generate those millions of test patterns.

GENERAL GUIDELINES USING MERCHANT DEVICES

Common to most devices covered in this chapter and to new devices, whether commercially available or custom-designed, are some general board level device testability implementation guidelines that cannot be overemphasized. These guidelines include the following:

- Providing control of clock lines
- Providing access to the control, address, and data busses (including internal busses)
- Providing access to the SYNC outputs or equivalent functions (e.g., address and data strobes)
- Providing access to the RESET, HOLD, WAIT, SINGLE-STEP, and INTERRUPT lines
- Providing means to tristate all devices
- Providing pull-up resistors on all tristate busses
- Providing control of CHIP SELECT and OUTPUT ENABLE lines (as well as direction lines on bidirectional buffer circuits)
- Partitioning static devices from clock circuits
- Partitioning analog circuitry sections

The sections that follow should be looked upon as a "library" of different circuit types, from the oldest to the newest. When faced with the task of selecting control and visibility points (in addition to address, data, and control busses) for a new device, try to find a similar one in the following sections for guidance in selecting the proper testability points for the new device. The general guidelines are summarized in Figure 6-1.

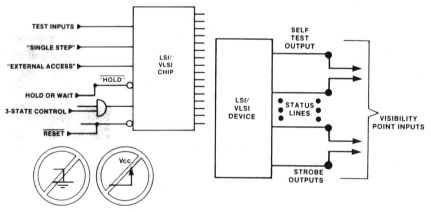

FIGURE 6-1.   Summary of general merchant device guidelines. Unused, from a functional standpoint, inputs on LSI/VLSI devices should not be tied hard to power or ground. Unused outputs also make valuable visibility points.

THE 8080A MICROPROCESSOR FAMILY

The 8080A family of microprocessor chips was one of the most widely used microprocessors several years ago. It is covered here because the control and visibility point requirements are unique and provide a baseline to show the evolution of device designs from typically difficult to test to much more testable today.

The 8080A (see Figure 6-2 for its pinout) is an 8-bit microprocessor which uses an external two-phase clock. The clock is divided into nine segments. The most common way to generate the required clock is with the 8224 clock generator (see Figure 6-3).

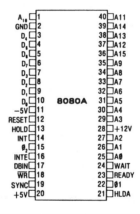

FIGURE 6-2.   8080A Pinout. It is quite simple to determine the correct control and visibility points for a device by looking at its pinout and reading the descriptions of the pin functions.

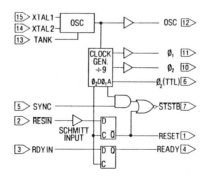

FIGURE 6-3.   8224 Block diagram. Another way to determine control and visibility points is to treat a device as if it were a board.

The 8224 uses a crystal and/or tank circuit to generate a signal (OSC) at nine times the 8080A clock frequency. A synchronization signal [ϕ_2(TTL)] is available for use with an ATE system that will execute test patterns at full speed (approximately 2 MHz). From a testability standpoint, since the 8224 cannot be initialized, it is highly desirable to provide tester control of the clock inputs to the 8080A. This may be accomplished in several ways. Figure 6-4 illustrates one method for isolating the clock lines.

In this example, control point A would be driven low by the ATE, and the phase 1 and phase 2 signals would be supplied by the ATE to control points B and C. Regardless of which clocking/ATE synchronization method is used, reset must be performed whenever the 8080A is powered-up. This reset operation must cover a minimum of three clock periods.

The reset operation should be performed first in all cases in order to initialize the contents of all internal registers (excluding status flags). Program execution will then begin at memory location 0000 (hex). One of the best testing strategies is to hold RESET active, which tristates the address and data busses, and let the ATE read and write (as applicable) the contents of ROM and RAM. This not only verifies the integrity of the busses but allows use of the 8080A as a source of test patterns using ROM and RAM after they have been proven to be good.

With control of the CHIP SELECT lines, the ROM at 0000 can be disabled following the reset operation and the ATE can emulate the ROM. In this way, a few instructions supplied by the tester can allow the

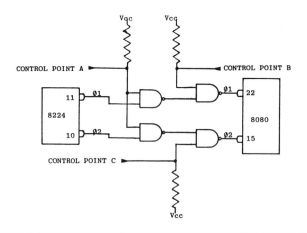

FIGURE 6-4.   8080 System clock control. Clock control, especially where two-phase clocks are concerned, often requires the addition of extra gates with controllable inputs.

8080A (in conjunction with RAM) to test itself. Enough instructions should be used to verify all internal registers and, for example, to make sure the program counter can count all the way up.

An exhaustive test of the 8080A, or any other sophisticated LSI/VLSI device, is not recommended at the board level. The testing objective is to make sure that the device was installed correctly and still functions after wave solder and touch-up. Parametric tests should be done at the chip level in receiving inspection.

The HOLD signal can also be used for the purposes already described. Its use provides the additional advantage of allowing the ATE system to single-step the microprocessor, instruction by instruction, at the ATE system's pin electronics rate, by alternating instructions with HOLD signals.

In the HOLD signal approach, with the ATE system emulating low-order address ROM, a test program can be written to allow the microprocessor to check each peripheral device on the board. Diagnostics by way of address failing or test step failing can be built into the test program to point to each device if this partitioning method is used.

Evaluation of data output by the microprocessor requires that the ATE system strobe only when data are valid. One way to derive the STROBE signal is with a circuit like that in Figure 6-5. Note that implementing this circuit in an interface requires visibility to the WR*, WAIT*, SYNC*, and ϕ_2 (TTL) lines.

The major testability guidelines for the 8224 clock generator and driver (pinout in Figure 6-6) are that access to the RESIN* be available and that the RESET output be connected to the microprocessor via a control point capability as illustrated in Figure 6-7.

With this method, independent testing of the 8224 can be performed. However, the internal divide-by-9 counter is not initialized with RESIN*. In fact, it is not initializable at all.

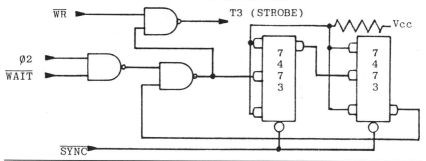

FIGURE 6-5. Tester strobe decoding for 8080 systems. Determining when data are valid on address and data busses sometimes requires extra circuitry, either on-board or in the test fixture.

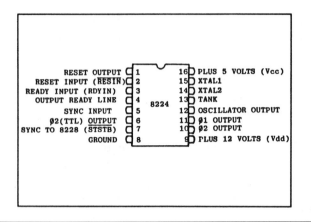

FIGURE 6-6. 8224 Pinout. Control is needed at the SYNC input to this device, and visibility is required at the phase 2 (TTL) output in order to synchronize a tester to the clock generated by the 8224.

Another device commonly used in 8080A-based systems is the 8228 system controller and bus driver. The pinout is shown in Figure 6-8. The key testability guidelines with this device are

A. Access to the BUSEN* pin, which allows the data bus I/O lines to be tristated

B. Access to the DBIN, HLDA, and WR* lines so that the test system can control the 8228 functions that will be transmitted to the system data and control busses

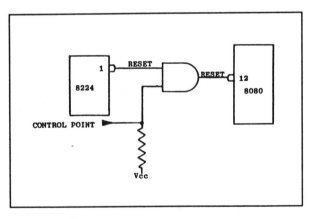

FIGURE 6-7. 8224 Reset control. When a reset signal is generated by a clock circuit (or other support device), it is important to have direct control of the devices being driven by it.

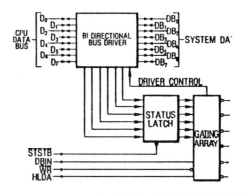

FIGURE 6-8.   8228 Diagram. System controller chips can be used to partition a board design as long as tester control of lines like bus enable (BUSEN*) and status strobe (STSTB*) are controllable.

One other device, with use not restricted specifically to 8080A systems (it is also used with the 8088, 8086, 80186, 80286, and 80386 processors), is the 8259 priority interrupt control unit (PICU). It is used for handling up to eight external interrupts and can be cascaded to handle up to 64 levels. The pinout and access points required for this device are shown in Figure 6-9.

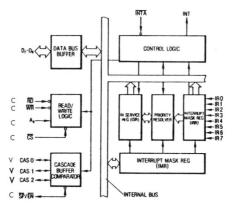

FIGURE 6-9.   8259 Diagram. Control of chip selects, which in this example include the A0 line and the lines that control bidirectional I/O lines, are important testability considerations.

THE 8085A MICROPROCESSOR FAMILY

The 8085A microprocessor differs from the 8080A in that

A.   The +12-V and −5-V power supply requirements are eliminated.

B.   The 8085A multiplexes its low-order address lines (A0–A7) with the data bus (D0–D7). The pins are labeled AD0–AD7 (see Figure 6-10).

C.   An on-chip clock, which may be driven by a single clock input that is twice the processor operating frequency, has been provided.

From a testing and testability standpoint, items B and C represent a mixed blessing. While the multiplexing of address and data lines complicates the ATE programming task, the ability to drive the on-chip clock from the ATE, as long as testability is considered, is a distinct advantage.

To design testability into a board using an 8085 (or similar

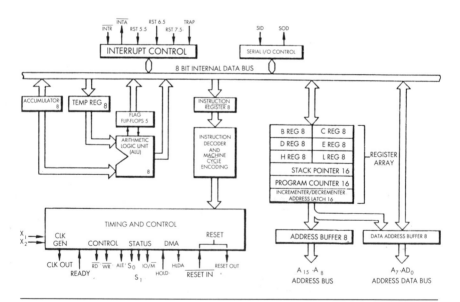

FIGURE 6-10.   8085 Processor block diagram. The RESET line for a processor often does not initialize all of the circuitry within the device. One of the first operations to be performed by a tester after releasing the RESET line is software initialization of the non-hardware-controlled circuitry.

processor), the first step is to provide for clock control. The 8085A on-chip clock can be supplied in one of three ways: with a crystal connected between the X1 and X2 inputs, with a resistor-capacitor network connected between those same inputs, or by driving the X1 (for low-speed operation) and X2 (for high-speed operation) inputs from an external clock source.

If the clock is generated with either the crystal or resistor-capacitor network, the test system will not have control of the clock lines. For testing purposes, the best method of providing clock control if either of these methods is used is to implement a jumper or a switch to allow the X1 input to be driven externally for test purposes (as if it were in the slave mode).

Just as important as clock control is access to control and visibility points. Edge connector access is best, followed by a special test connector, IC clips, or a bed-of-nails fixture. With the 8085A, the control points needed include the READY, HOLD, TRAP, and RESET IN* lines. Access to these lines provides for initialization of the microprocessor, tristate of address and address/data lines to facilitate a partitioned test programming approach, interleaving test program data with HOLD signal functions on I/O rate ATE systems, and control of the nonmaskable interrupt (TRAP).

The test system must control the TRAP line in order to prevent a condition on the PCB under test from causing response to an interrupt when there may be uninitialized data internal to the 8085A. Should an interrupt be serviced with the processor not totally initialized, unknown or "X" states will be propagated throughout the unit under test.

The visibility points that should be available to the ATE system include

- The ADDRESS LATCH ENABLE (ALE) line; this signal can be used in place of the SYNC signal of the 8080A.
- The S0 and S1 bus state indicator lines; these status signals can be used by the tester (or tester interface) to decode the type of operation being performed at a given time.
- The RD*, WD*, and IO/M* lines; these signals can be used to determine proper clocking, strobing, and data availability to and from the unit under test.

Some of the above lines go into the tristate condition in response to control signals. All lines that can be tristated, on this or any other processor, should have pull-up resistors attached (Figure 6-11). The pull-up resistors, as shown in the preferred method, should be placed within the board boundary so that there is a determinable logic state (as

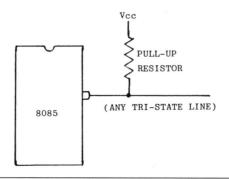

FIGURE 6-11.   Pull-up resistors on tristate lines. Placing pull-up resistors on tristate lines, either on-board or in the test fixture, provide nice, clean sharp edges and prevent false errors from being flagged.

opposed to a floating output) when the lines of the active device are in the tristate condition.

The 8085A is normally used with different support devices, unless the low-order address lines are not used or are demultiplexed by external logic, in which case the 8080A support chips may be used. Some of the devices used specifically with the 8085A include the 8155 static read/write memory with I/O ports and timer and the 8355 ROM with I/O.

The 8155 block diagram, annotated to show the needed board level control and visibility points, is shown in Figure 6-12. The 8355 block diagram, similarly annotated, is shown in Figure 6-13.

The reset operation does not clear the timer, the internal memory,

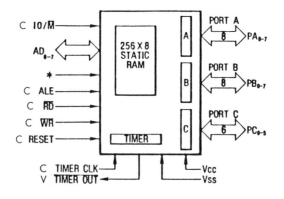

FIGURE 6-12.   8155 Diagram. Control of RAM devices centers around the RESET, READ, WRITE, and ADDRESS LATCH ENABLE lines.

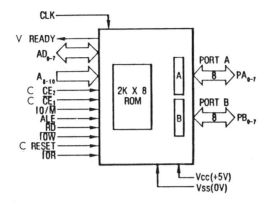

FIGURE 6-13. 8355 Diagram. For ROM devices, the major testability guideline is to provide control of the CHIP SELECT lines. When more than one line is available, one line can be used for functionality and the other for testability.

or I/O locations within the 8155 device. To fully initialize this device, specific software initialization must be performed. If, for example, the I/O port addresses are C0 through C5, an appropriate 2-byte initialization sequence would be

MVI	5A, 80
OUT	5OC 4
MVI	5A, 60
OUT	5OC 5

The static RAM portion of the device should also be initialized and verified before letting the 8085A use it for testing purposes. This is normally accomplished by writing a specific code to all memory locations, followed by a read to verify the correctness of the write operation. Alternatively, the RAM portion of the 8155 can be tested with a specific pattern (such as checkerboard) on a partitioned basis and then initialized for later use by the unit under test when it runs as a system.

The 8355 ROM is 2048 bytes by 8 bits of ROM and has two I/O ports. This device allows reading of memory and I/O functions between either of the two ports and memory or from the system data bus. As illustrated, the key testability guideline is access to control lines, so the device can be

A. Tested by itself as an entity
B. Deselected and tristated so that other portions of the PCB under test can be exercised

Another commonly used device, the 8755 erasable programmable read only memory (EPROM) with I/O, has minor signal and pin differences from the 8355A. In addition, it is erasable with ultraviolet light and may be reprogrammed. The testability considerations for use of this device are the same as those for the 8355 ROM.

For either the 8355 or 8755A, or any other ROM, PROM, or EPROM for that matter, the contents of memory should be read from every address and be verified with either a 1's and a 0's stored pattern, a signature, or a checksum before allowing the system microprocessor to access the data. This is especially true for a ROM containing self-test and diagnostic routines.

THE 8048 MICROPROCESSOR FAMILY

This family of devices is more aptly named a family of microcomputers, rather than microprocessors, since each device contains its own ROM and scratch pad memory and thus, with a single chip, can operate as a complete system for simple applications.

The devices in this family include the 8048, 8049, 8748, 8749, 8035, and 8039. The pinout and block diagram for the 8048 (and its cousins) are depicted in Figure 6-14. Note that it very much resembles the general case of a microprocessor-based printed circuit board.

The typical testing procedure for these devices would include initial reset, verifying internal ROM, initializing and/or loading scratch pad memory, and allowing the processor to execute the testing patterns.

The key testability considerations are aimed at control and visibility. Control points required include

- T0, the test input
- EA, the external program memory access line
- SS*, the single-step control line
- INT*, the interrupt request
- RESET*, the chip reset line

Visibility points required include

- ALE, the address latch enable (SYNC) line
- RD*, the data memory read control
- WR*, the data memory write control
- PSEN*, the external memory read control line

Since these devices contain on-chip clocks, the same technique mentioned for the 8085A should be employed for clock control, unless the ATE system can synchronize with the crystal or RC network and keep up with the microcomputer. Strobe synchronization should be referenced to the ALE line.

After the reset operation, the T0 line should be used to verify internal memory on the device. The EA line can be used to force the device into the debug mode, where it must access external program memory—in this case, an ATE system.

These devices also have a single-step mode to allow relatively simple interface requirements and to make testing on I/O rate test systems less burdensome. A typical circuit configuration which allows the ATE system to control and use the single-stepping feature is shown in Figure 6-15.

With this circuit, which can be implemented either in hardware in the test fixture or by programming the ATE system to emulate the hardware function performed by the circuit, the line marked "MODE", which comes from the ATE system, determines whether the processor will run at speed or in the single-step mode (i.e., instruction by instruction). With a logic 0 applied to the MODE input, the processor will run.

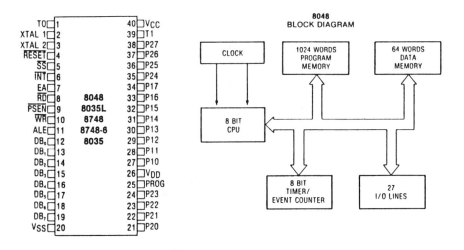

FIGURE 6-14.   8048 Pinout and block diagram. Microcomputers have on-ship ROM which must be disabled for test purposes if the automatic test equipment is to be able to supply the processor with test vector data that will not fit in the on-chip ROM.

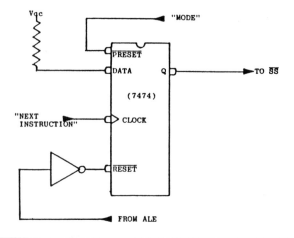

FIGURE 6-15.   Single-stepping with hardware. The ability to force a processor to execute one instruction at a time is important when the tester data rates are not as fast as the data rates of the device under test. Hardware is sometimes required to create a single-step mode.

A logic 1 applied to this input will force the single-step mode, and each instruction will execute on activation of the "NEXT INSTRUCTION" line, which also comes from the ATE system.

This suggested circuit can be easily added to a PCB under test. It may also be used by the designer to aid in debugging when using standard development tools. It is a plus for both the test engineer and the design engineer. Whether or not SS* is used in the normal circuit operation, it should always be accessible for testing purposes. It should not be tied directly to the V_{cc} line on the circuit board.

Another important line on devices like the 8048 that contain on-chip ROM is the EA* (external access) line. This line allows the on-chip ROM to be disabled and forces the processor to access external memory (e.g., an ATE system) for debug and test purposes. Lines that perform EA* functions should also be made controllable at the PCB test level.

THE 8086 MICROPROCESSOR FAMILY

The 8086 is a 16-bit microprocessor. In addition to the simple increase in bus size, there are some internal changes between the 8086 and its predecessors which cause increased test complexity. Some of these changes and additions include the following:

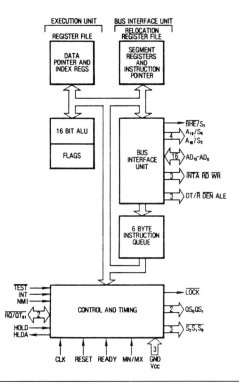

FIGURE 6-16.   8086 Block diagram. Devices with multiple processors on-chip often have a line (like the TEST* line on the 8086) that can be used to synchronize the operation of both portions of the device.

A.   The 8086, designed to work in large and small configurations, has several output pins which may have different signals depending on the state of pin 33 (MN/MX*). These different signals, depending on the selection of "minimum system" or "maximum system" usage, are listed on the block diagram of the 8086 in Figure 6-16.

B.   The CPU has been divided into two halves, called the execution unit (EU) and the bus interface unit (BIU). These two sections operate asynchronously.

C.   The 8086 can have its own local memory and can also share common system memory in multiple processor designs. This means that particular care must be taken in system (board) design to insure control of all memory elements.

The 8086 uses a multiplexed address/data bus, which can complicate both the test interfacing and the test programming from a timing
and handshake standpoint. It is one of the few processor designs,
however, that incorporates at least minimal design for testability in the
form of pin 23 (TEST*).

The TEST* pin can be used for various purposes, including the
following:

A.   The wait state for the execution unit can be initiated via
     software (as opposed to the hardware wait state, which is
     forced by the READY input) to cause a continuous sequence
     of idle clock periods. The BIU is still active, however, and
     can be used to fill up the instruction queue via memory read
     bus cycles, thus allowing software initialization of the EU.
     The wait state is terminated by applying a logic 0 to TEST*.

To use this capability from a testability standpoint, the test engineer
must initiate a program-induced wait state immediately following the
reset operation. The design can allow this access in various ways, including socketing the ROM devices or providing chip deselect control.
This way the test programmer can use the ATE stimulus patterns to
initiate the programmed wait state and achieve software initialization
on the board.

B.   The wait state can be used to synchronize two or more
     processors in a multiple processor system or to synchronize
     the processor with the ATE system. Thus the EU can be
     loaded (via the BIU) with simple subroutines, whose
     execution can be controlled via the TEST* pin. Execution
     will begin when TEST* is driven low.

As with all microprocessors, the reset operation should be performed first. The 8086 has an asynchronous RESET input which can be
driven high at any time. The logic 1 state must remain for at least four
clock cycles, and the reset operation begins when the RESET pin is
driven to logic 0. The reset operation takes approximately 10 clock
periods for completion and must be synchronous with the clock when
RESET is driven low.

An example of an 8086 testing strategy might be to reset the device,
input a WAIT instruction from location FFFF0 (e.g., from the ATE
system), and then exercise each portion of the device with a simple
program such as

A.   Drive TEST* to logic 1

    B.   Input the first test: wait, instruction, data, instruction, data, (tests the program counter)

    C.   Pulse TEST to logic '0'

    D.   Check result

    E.   Input the next test: wait, instruction, data, instruction, data,, (tests the ALU)

The ATE system can then cause each small subroutine test to execute by pulsing TEST* to a logic 0 and checking the results when the EU has entered the next wait state.

    The control points needed for testability of the 8086 include

- NMI, the nonmaskable interrupt request line
- INTR, the interrupt request line (maskable)
- CLK, the system clock, derived from the ATE or, more commonly, from an 8284 clock generator/driver
- TEST*, the test input line
- RESET, the device reset line
- READY, the wait state request line

Visibility points needed, in addition to the address/data bus lines, include

- BHE*/S7, the high-order byte/status output
- RD*, the read control line (to allow handshake with the ATE and recognition of proper clock and strobe times)
- Pins 24 through 29, the multipurpose outputs

Their actual use by the ATE system will depend on the state of the MN/MX* line. Most of these visibility points can be tristated to allow testing of other devices on the board by using the previously described partitioning approach.

    The 8086 derives its clock input from the 8284 clock generator/driver. The block diagram for this device is shown in Figure 6-17. The control points needed for this device include

- CSYNC, the clock synchronization line
- EFI, the external clock input
- F/C*, the clock source select line
- RES*, the reset input (which is later transformed to the 8086 reset output at pin 10, RESET)
- RDY1 and RDY2, the wait state inputs
- AEN1* and AEN2*, the address enable qualifiers for RDY1 and RDY2

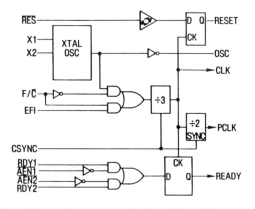

FIGURE 6-17. 8284 Block diagram. A clock circuit that includes the on-chip circuitry for allowing an external clock source input is very desirable from a testability standpoint.

In contrast to the techniques described earlier for initialization, control, and synchronization of the 8224 (which can also be applied to this device), the 8284 can be more simply driven via the EFI input from the ATE. The input at EFI must be three times the frequency of the 8086 clock period. Provision must be made on the board under test to disconnect the tank and/or crystal circuits via the F/C* line in order for the ATE system to control the 8284 frequency.

Since the 8086 is designed for multiple processor applications, it is important that all clocks be derived from a single source and not from multiple asynchronous 8284 devices. And the master clock should not have its EFI and F/C* lines tied off hard to power and ground lines. Using slave 8284s and having the master 8284 controllable externally results in a very testable clock circuit implementation. Thus the 8284 is a preferred device selection from a testability standpoint.

The completion of an 8086-based system requires the use of the 8288 bus controller. The block diagram for this device is shown in Figure 6-18. All of the 8288 control outputs normally connected to the system bus can be tristated as long as control of the IOB (mode control) line is possible. If the system design uses the 8288 in the I/O bus mode, the IOB pin should not be tied directly to V_{cc}. It should be tied through a pull-up resistor to allow the test system to isolate the 8288 from the bus.

When a microprocessor board is to be tested running at speed, the 8288 will provide an output, AMWC* (advanced memory write con-

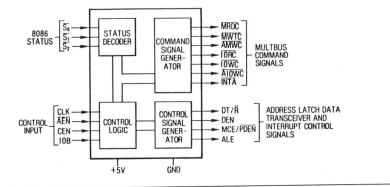

FIGURE 6-18.   8288 Block diagram. Control points can readily discerned as the inputs to the block called Control Logic in this (and similar) devices, while visibility points are chosen from the outputs of the Command Signal Generator.

trol), which is one clock pulse in advance of the normal 8086 write signal. This will allow setup of ATE response data in advance of the 8086's output of it, thus simplifying the strobing requirements.

Other input control points where access is desirable include

- CLK, the clock input
- AEN*, the bus priority enable/control line
- CEN, the control enable line

The proper control input states for isolating the 8288 from the rest of the system are

AEN*	CEN	IOB
1	0	0

Two other devices that support the 8086 are the 8282 eight-bit I/O port and the 8286 eight-bit bidirectional bus transceiver. For these devices, access is needed to the OE* and STB pins (8282) and to the OE* and T pins (8286).

A great many other support devices are used with the 8086. For the most part, the guidelines for these devices follow those described for the devices covered so far.

THE 80186 PROCESSOR

Like the 8086, the 80186 is composed of independent bus interface and execution units. It also has two independent high-speed direct memory access (DMA) channels and three programmable 16-bit timers along with a programmable wait state generator and a local bus controller integrated on-chip. The block diagram for the 80186 is illustrated in Figure 6-19.

Control of the system reset line (RES*) is obviously required in order to initialize the processor after power is applied. A system reset causes the 80186 to immediately terminate its present activity, clear the internal logic, and enter a quiescent state. The 80186 will begin fetching instructions approximately seven clock cycles after the RES* line is returned to the logic 1 value. Thus it is important to have initial instructions set up prior to releasing the RES* line.

The RES* line must be held low for a minimum of four clock cycles, and may be applied asynchronously because it is internally synchro-

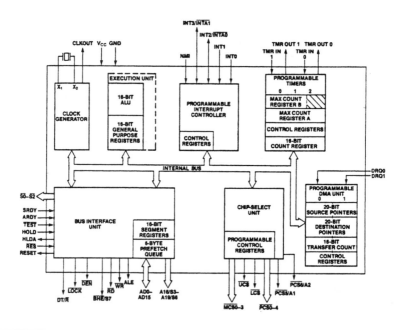

FIGURE 6-19.   80186 Block diagram. As processors become more complex, the number of control and visibility points required to make the boards they are used on testable increases. The primary control and visibility points required are those going to and from the Bus Interface Unit block.

nized by the 80186. For proper initialization, the low-to-high transition of the RES* line must occur no sooner than 50 microseconds after power is applied. When RES* goes low, the processor will drive the status lines to an inactive level for one clock period and then tristate them.

Control is also needed for the TEST* line. This line operates in conjunction with the WAIT instruction of the processor. If the TEST* line is at the logic 1 state when a WAIT instruction is executed, instruction execution will be suspended. The state of the test line will be continuously examined until it goes to the logic 0 state, at which time execution will resume.

Another line needing control is the HOLD input. This line tells the processor that another bus master (e.g., a piece of automatic test equipment) is requesting control of the local bus. HOLD, which is an active high signal, may also be asserted asynchronously with respect to the 80186 clock. When the processor acknowledges the HOLD signal, it will signal that acknowledgment by asserting the HLDA line and by tristating the local bus and control lines. This is important if it is desired to test other devices on the bus without having to use the processor to do so.

Some 80186 systems use the ARDY (asynchronous ready) or SRDY (synchronous ready) lines to inform the processor that either addressed memory space or an I/O device will complete a data transfer. One of these signals must be used, and the Intel data book says to tie the other signal to ground if unused. For testability purposes, the unused signal should be tied to an controllable inverter with a pull-up resistor at its input, and the active signal should be gated such that it is also controllable by the ATE.

To execute operations on the DMA channels of the 80186, the DRQ0 and DRQ1 lines must be controlled. The standard guidelines for clock control and external masking of NMI also apply to this device.

Visibility points required, in addition to the address and data busses, include CLKOUT (in order to synchronize with the clock if it is not externally supplied by an otherwise controllable clock source), S0*–S2* (in order to decode bus transaction information), and ALE*, WR*, and RD* (which also double as status outputs QS0–QS2).

Finally, it is also helpful to have visibility to the DT/R* and DEN* lines. These lines control data flow through external data bus transceivers. When DT/R* is low, the data on the bus are being transferred into the 80186. Conversely, when the line is high, the 80186 is placing data on the bus. The enable and direction lines of the chips being driven by DT/R* and DEN* should be gated such that an ATE can substitute its own enable and direction control signals for the processor signals when the processor is in the WAIT or HOLD mode.

THE 80286 PROCESSOR

The 80286 is a high-performance VLSI microprocessor that supports both multiuser reprogrammable and real-time multitasking operations. A typical 80286 system, as illustrated in Figure 6-20, includes not only the processor itself but also several support components (i.e., the 8259A programmable interrupt controller, the 82C284 clock generator and driver, the 82288 bus controller, the 82289 bus arbiter, and various latches and buffers).

The control points needed for the 80286 are the same as those needed for the 80186. The locations of some of those control points are different, however. The ARDY* and SRDY* lines, for example, have been moved to the 82C284 clock generator chip. These signals are synchronized inside the 82C284 and then sent to the 80286 via the READY* line.

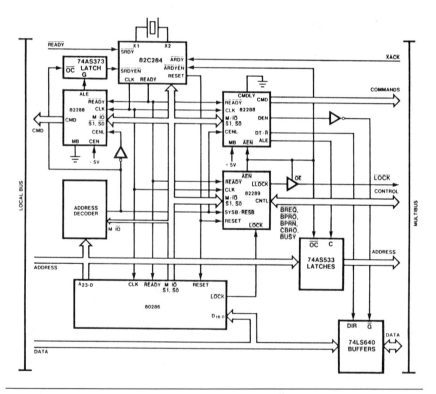

FIGURE 6-20. Example 80286 system block diagram. A typical processor-based board design needs control not only of the processor but also of the clock chip, address decoder, buffers and latches, bus arbiter, and bus controller.

The example system in Figure 6-20 needs to be improved from a testability standpoint according to the guidelines set forth in Chapter 5. Visibility and control are needed for the address and data busses, as well as the command signals of the 82288.

External clock control for the 82C284 is desirable. If it cannot be implemented, visibility is needed to the CLK output in order to synchronize to the processor system. Control also needs to be provided for the OC* and G inputs to the 74AS373 latches in order to provide a tester input to the 82C284 SRDYEN* line (which works with the SRDY* line, already mentioned as a needed control point).

The MB and CEN lines on the left-hand 82288 are tied hard to ground and +5 V, respectively. The MB line should be tied low via an inverter with a pull-up resistor at its input, and the CEN line should have a pull-up resistor so that the chip can be disabled without requiring extra external logic. Alternatively, the inverter driving the CENL line could be changed to a two-input gate, which would also provide chip select control for the 82288.

The OC* and C lines of the 74AS533 latches and the DIR and G* lines for the 74LS640 buffers should be gated so that tester control of these signals is possible. Another option is to replace the 74AS533 latches with a part containing bidirectional latches. This would allow access to the internal address bus from the address lines on the right side.

Visibility points needed include the M/IO* line and S0 and S1 status output lines. Note that not all of the pins on the 80286 are shown in Figure 6-20. This is quite typical, and care should be taken to make sure that every pin on the processor itself that needs control is indeed controllable.

THE 80386 PROCESSOR

The internal architecture of the 80386 consists of six functional units that operate in parallel. Fetching, decoding, execution, memory management, and bus addresses for several instructions are performed simultaneously. The six functional units that make up the pipelined instruction processing 80386 are the bus interface unit, the code prefetch unit, the instruction decode unit, the execution unit, the segmentation unit, and the paging unit.

The execution unit in turn consists of three subunits: the control unit, the data unit, and the protection test unit. Figure 6-21 illustrates the interconnection of these functional units in the 80386, while Table 6-1 summarizes its signal pins.

The control points needed for the 80386 include the

- CLK2 input (either directly or indirectly)
- HOLD line
- READY# line
- RESET line
- PEREQ (coprocessor request) line
- NA# (next address request) line
- BUSY# line

Visibility points, in addition to the standard requirements, include the

- D/C# (data-control indication) line
- W/R# (write-read indication) line
- M/IO# (memory-I/O indication) line
- ADS# (address status) line

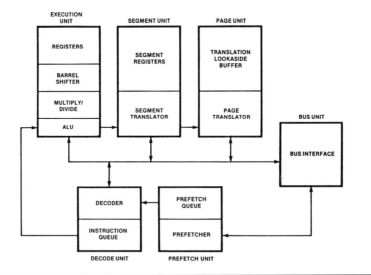

FIGURE 6-21.   80386 Functional units. Processors with multiple internal units that run in parallel can be especially challenging to test. Signals to the processors multiple internal blocks must be controlled via the bus interface unit.

TABLE 6-1. Summary of 80386 Signal Pins

Signal Name	Signal Function	Active State	Input/ Output	Input Synch or Asynch to CLK2	Output High Impedence During HLDA?
CLK2	Clock	—	I	—	—
D0–D31	Data bus	High	I/O	S	Yes
BE0 # – BE3 #	Byte enables	Low	O	—	Yes
A2–A31	Address bus	High	O	—	Yes
W/R #	Write-read indication	High	O	—	Yes
D/C #	Data-control indication	High	O	—	Yes
M/IO #	Memory-I/O indication	High	O	—	Yes
LOCK #	Bus lock indication	Low	O	—	Yes
ADS #	Address status	Low	O	—	Yes
NA #	Next address request	Low	I	S	—
BS16 #	Bus size 16	Low	I	S	—
READY #	Transfer acknowledge	Low	I	S	—
HOLD	Bus hold request	High	I	S	—
HLDA	Bus hold acknowledge	High	O	—	No
PEREQ	Coprocessor request	High	I	A	—
BUSY #	Coprocessor busy	Low	I	A	—
ERROR #	Coprocessor error	Low	I	A	—
INTR	Maskable interrupt request	High	I	A	—
NMI	Nonmaskable intrpt request	High	I	A	—
RESET	Reset	High	I	S	—

The 80386 contains built-in test features that improve its testability considerably when compared to earlier devices and even current devices from other manufacturers, both at the chip and board levels. The 80386 testability features are centered around both signature analysis and Intel proprietary test techniques. All of the regular logic blocks in the chip, which make up about 50 percent of all its internal devices, can be tested by using built-in test features.

Features have been included in the 80386 to support two types of tests: automatic self-test and translation lookaside buffer (TLB) tests. The built-in automatic self-test is completely contained in the processor, and all that is needed to perform it is to initiate it and check the results. For TLB tests, which are externally developed and applied, the 80386 provides an interface to simplify the process.

The automatic self-test is initiated by activating the BUSY# input during the initialization of the processor. To initialize the 80386, the RESET input must remain high for at least 15 CLK2 periods. When self-test is to be performed, the RESET line must be held high for at least 80 CLK2 periods.

Initialization begins based on the rising edge of the signal on the RESET line. The RESET line should be held high for at least 1 millisecond after V_{cc} and CLK2 are up to voltage and speed. On the falling edge of the RESET signal, the BUSY# signal is sampled. If BUSY# is low at that time, the processor will perform a self-test whose results are stored in the EAX register, which may be externally interrogated.

The on-chip self-test checks the three major programmable logic arrays (i.e., the entry point, control, and test PLAs) and the contents of the control ROM. It uses the linear feedback shift register technique (see the second section of Chapter 7 for a discussion of this topic) to generate test vectors and take a signature of the resulting response vectors. This signature is compared on-chip with a signature constant stored in the 80386. If the signature register contents match the signature constant, the result of the comparison will be all zeros, which are then loaded into the EAX register.

The 80386 continues with the initialization process started by driving the RESET line high regardless of the results of the self-test. Thus it is important that the software (or firmware, as the case may be) in the system check the self-test results as one of its first tasks.

For externally applied TLB tests, the 80386 has two 32-bit registers that can be written to and read from using the MOVE TREG, reg and MOV reg, TREG instructions. Test register 6 is used as the command register, which handles both addresses and commands. Test register 7 is used as a data register and can be written to or read from under control of an automatic tester or assembly language test programs residing in external ROM.

THE Z80 MICROPROCESSOR FAMILY

The Z80 is similar to an 8080A three-chip microcomputer system in that it contains the functions of CPU, clock (8224), and system controller (8228). It requires only a single +5-V power source and a single clock input. The pinout is shown in Figure 6-22.

Z80 control and test point requirements are as follows:

Control Points	Test Points
RESET*	RD*
BUSRQ*	WR*
WAIT*	IORQ*
NMI*	M1*

As with all of the processors discussed thus far, provision for driving the clock input externally from the ATE system is an extremely important requirement.

All of the discussion of the 8080A family also applies to the Z80 device. Access to RESET* is required to isolate the CPU for a partitioned test and to reset the program counter, IV, and R registers to zero. Interrupts are disabled after a reset operation except for nonmaskable interrupts via NMI*. Design criteria should insure that the test system can prevent or control NMI* so that no nonmaskable interrupts occur or are serviced prior to software initialization.

Access to WAIT* allows an I/O rate test system or in-circuit test system to be used to test the Z80. The BUSRQ* line allows tristating of

FIGURE 6-22.   Z80 Pinout. Control and visibility lines for this (and similar) devices center around clocking, initialization, halting, and waiting while monitoring READ, WRITE, and I/O REQUEST lines.

the address and data busses for access to other devices without resetting the Z80 (or the entire printed circuit board under test).

The output test points (RD*, WR*, and IORQ*) are needed to allow the interface to decode proper tester clock and strobe times (in conjunction with the clock input signal to the Z80), so only valid data are driven in to the device and output data are valid when compared.

To test the interrupt circuit, it is often desirable to have a signal available called INTERRUPT ACKNOWLEDGE. While there is no single Z80 pin with this function, the function can be derived by ANDing (watching out for proper logic signal polarities) the IORQ* and M1* signals.

Many of the 8080A support devices can be used in Z80 systems. Two devices, the parallel I/O (PIO) interface and the counter/timer circuit (CTC), are unique to Z80 designs. The Z80 PIO is a 40-pin DIP which requires a single +5-V power supply and a single clock input. Since in most systems this clock input will be driven from the same clock as the Z80, separate access is not required (assuming that the guideline for Z80 clock control has been followed).

Important on-board control points for this device include

- CE*, the chip enable line. Access to this line is required to tristate its outputs for testing other devices on the bus.
- B/A* and C/D*, the port select and control versus data select lines. Access to these lines provides the means for a thorough device test without using the Z80 CPU to generate stimulus test vectors.

Important visibility test points for this device include

- A RDY and B RDY, the port A and port B ready lines. These provide information to the test system for clocking and strobing data in and out of the PIO.
- INT*, the interrupt request output. This output goes low when an interrupt is to be output to the Z80 CPU. Note that this output is the complementary metal oxide semiconductor (CMOS) equivalent (open drain) to a TTL open collector line and should therefore have a pull-up resistor provided so that it will not float in the "unknown" or "error" region. The PIO can be reset by simultaneously inputting a logic 0 on MI* and logic 1's on IORQ* and RD*.

In order to test the Z80 PIO, it must first be set up with a control code. The format for appropriate control codes is shown in Figure 6-23. The Z80 CTC is a programmable device with four sets of timing logic which are separately programmable as interval timers or external event counters.

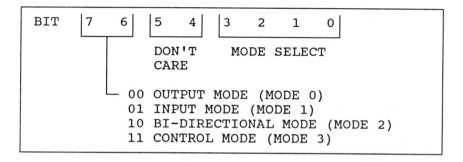

FIGURE 6-23.   Z80 PIO control codes. It is often necessary to configure a device in order to test it. This software initialization often requires reference to the programming instructions for the device (in addition to its block diagram and signal pin descriptions).

Test and control points for this device are similar to those described for the PIO and are summarized as follows:

Control Points	Test Points
CE*	A*D INT*
RESET	(A*D open drain)

Testing this device requires software initialization (if done by the processor) or its equivalent via stimulus from the ATE system. The procedure is as follows:

- Output an interrupt vector when initializing the CTC
- Output control codes to each of the four sections of the device

```
LD      A,      2C
LD      1,      A
IM      2
LD      A,      40
OUT     (0B8),  A       (INTERRUPT VECTOR)
LD      A,      0C5
OUT     (0B8),  A       (CONTROL CODE)
LD      A       DATA    (DATA=INITIAL COUNT)
OUT     (0B8),  A       (START CHANNEL 0)
 .       .       .
 .       .       .
 .       .       .
```

FIGURE 6-24.   Z80 CTC control code example. Testing of some devices requires that actual assembly language code by developed and then converted to 1's and 0's for the tester's pin electronics cards.

The interrupt vector specifies which channel will receive the control code. An appropriate sequence for channel 0 as a counter via the Z80 CPU is shown in Figure 6-24.

THE Z8000 MICROPROCESSOR FAMILY

The Z8000 family of processors includes the Z8001 and Z8002 CPUs, the Z8010 memory management unit (MMU), and the Z8034 universal peripheral controller (UPC).

The Z8001 is a 48-pin DIP, and the Z8002 is a 40-pin package. The primary difference between the two devices is that the Z8001 has the extra pins to handle additional external memory segments. For testability purposes, this section deals with the Z8002.

Like the 8086A family, the Z8000 CPUs multiplex the address and data lines. Many of the comments applicable to the 8086 are also applicable to the Z8000 family.

The Z8000 requires a single clock input which should be controllable by the ATE. Alternatively, if a high-speed ATE system is used, it can synchronize to the clock on the board under test as long as access has been provided.

All but two of the output lines from the Z8000 can be tristated. Pull-up resistors, therefore, should be provided on these lines.

Output test points needed (in addition to the standard visibility rules) include

- R/W*, the read/write select lines
- ST0-ST3, the machine cycle status lines
- AS*, the address strobe line. This line is low when an address is being outputted by the processor and always occurs at clock cycle T1.
- DS*, the data strobe line. This line is low when data is being inputted or outputted to and from the processor and always occurs at clock cycle T3.
- MREQ*, the memory request line

Access to these lines will assist the test engineer in clocking and strobing data to and from the test system and the board under test. The address and data strobe lines provide the necessary information about the state of the bus so that the proper operation can be performed at each test step.

Input control points needed for Z8000 systems include

- RESET*, the reset input to the device. A low at this input will initialize the processor and tristate all outputs (except BUSAK* and MO*). Access to RESET* should be provided.
- BUSRQ*, the bus request input. Access to this line allows tristating of the address/data bus without resetting the CPU or UUT. This function for the Z8000 is analogous to the hold state in other processors.
- NMI*, the nonmaskable interrupt. If access is not provided, provision should be made to preclude occurrence of a nonmaskable interrupt before the test system can accomplish software initialization.
- WAIT*, the wait state input. This function will allow the CPU to insert idle clock periods into machine cycles to let the board be tested at less than full clock speed.
- STOP*, the stop input. This line allows control of the processor on an instruction by instruction basis (as opposed to cycle by cycle with the WAIT* line). This line provides a function similar to the one implemented for the 8048 family (single step).

The 8010 MMU is used with the Z8001 when very large memory requirements exist. The main consideration with this device is control point access to CS*, the chip select input, which allows isolating the device from others on the UUT.

Another device still encountered many times is the Z8034 UPC. This device is actually a microcomputer in its own right and operates as a slave to the Z8000. It has 2,048 bytes of on-chip ROM. The code in this ROM should allow immediate access to control of the device by the test system. The CS* line should be controllable by the ATE system.

THE 6800 MICROPROCESSOR FAMILY

The 6800 microprocessor is an 8-bit, single-power-supply (5-V) device with an external clock. It has perhaps the simplest timing considerations of any of the MOS microprocessors covered in this text. This timing simplicity occurs because a clock cycle and a machine cycle are one and the same for the 6800. Tristate control for the data and address busses of the 6800 has been split, with three-state control (TSC) providing for floating of the 16-bit address bus, and data bus enable (DBE) performing a similar function for the 8-bit data bus.

A third line, HALT*, floats the entire system bus (address, data, and the R/W* output) and causes the CPU to stop execution when the currently executing instruction has been completed.

One of the key guidelines, from both testing and system design standpoints, is that RESET must be held at logic 0 for at least eight clock cycles (both for power-up and during an in-process reset). From a testability standpoint, the preferred methods for accomplishing this are shown in Figure 6-25 in decreasing order of preference.

Method 1 is a compromise between method 2, which needs fewer parts but requires clip access (or edge connector access, in which case it is the preferred method), and method 3, which requires more components.

After a reset operation, the 6800 will load from hexadecimal addresses FFFE and FFFF. Provision should be made for allowing the CPU to access the ATE system at these memory locations so that the test system can control the initialization of the printed circuit board under test. This can be accomplished by socketing the PROM at address XXXX–FFFF or by providing access (via a control point) to the CHIP SELECT line for the PROM at address XXXX–FFFF so that the test system can deselect it and emulate locations FFFE and FFFF.

Output test points which are important for synchronization of data input and output include VMA and R/W*.

The following list summarizes the testability rules for the 6800 microprocessor.

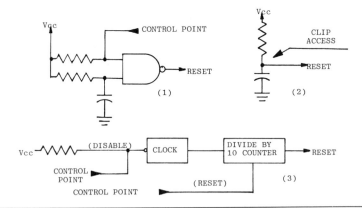

FIGURE 6-25.   6800 Reset methods. Some processors require that the RESET line be held in the active state for a certain number of clock periods. External tester control is needed, and synchronous methods are preferred over asynchronous methods.

6800 Testability Summary

A. Provide control of ϕ_1 and ϕ_2 clock inputs.
B. Provide input control points for RESET, NMI*, DBE, TSC, HALT*.
C. Provide output test points for R/W*, VMA, BA.
D. Provide for address and data bus access.

These rules conform quite closely to those given in the general guidelines section.

The 6802 CPU adds an on-chip clock and RAM to the 6800. The clock control rules discussed earlier still apply.

The address bus cannot be floated with this device even though HALT* will float the data bus. Proper design for testability (which will also allow the system design to address the address bus) is to buffer A0–A15 with drivers that have a tristate control.

Another key consideration is control of the RE line (on-chip RAM enable) so that the ATE system can check out the microprocessor functions (from the same memory locations as for the 6800) and then perform software initialization of the on-chip RAM.

An advantage to be gained here is that, after a write/verify with read sequence, a sequence can be loaded into RAM and used to let the CPU free-run to test other portions of its functions, thus reducing the test program stimulus generation effort.

Support devices for the 6800 family include clock logic devices (6870, 6871, 6875), peripheral interface adapters (6820 and 6520), and an asynchronous communication interface adapter (6852). Each is briefly discussed from a testability standpoint.

The 6870 is the least desirable choice, since it requires the most complex clock control circuitry (external to the chip, since there are no hold or reset functions). The ϕ_2(TTL) output can provide synchronization to a high-speed automatic test system.

The 6871 is a better choice, since it provides a hold line input (HOLD1*), memory stretching to allow use of a slower ATE, and a clock output twice the frequency of the UUT system clock (which, if accessible, can be used by the ATE and its interface to set up stimulus data clocking and response data comparison). Control of HOLD1* and memory clock, with access to 2XFE and ϕ_2(TTL), constitute the testability guidelines for this device.

The 6875 is the preferred device from both testability and system design standpoints. It allows external drive for synchronization to multiple clock generators and ATE plus an asynchronous reset input (SYS

RES*). An example of clock control with this device when an on-board crystal is used is shown in Figure 6-26.

The 6820 and 6520 peripheral interface adapters (PIAs) have identical pinouts and are representative of many devices. Test and control points for these devices (and others like them) are summarized as follows:

Control Point (input)	Test Point (output)
CS0, CS1, CSA*	IRQA*
RS0, RS1	IRQB*
	*E
	RESET*

Note: *E line access is not required if it is driven from a controllable system clock

Asynchronous communications interface adapters (ACIAs) are also quite popular devices. An example is the 6850 ACIA. This is a 24-pin DIP device whose control and test points include

Control Points	Test Points
CSO	RS
CS1	R/W*
CS2*	RxD
RxCLK	TxCLK
CTX*	DCD*

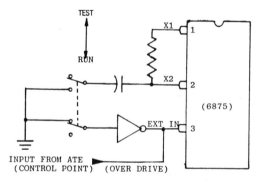

FIGURE 6-26.   Clock control for the 6875. A switch (or jumper) can often be used to control an on-chip clock oscillator and provide for an external input to the device from a tester.

One key use of this device on a printed circuit board is to provide for synchronization of the 6850 with the ATE system, even though the part has been designed to operate asynchronously. Another is to provide for software initialization, since the device has no reset input and requires a control code. The master reset control code for this device is

Bit	7	6	5	4	3	2	1	0
	X	X	X	X	X	X	1	1

Synchronous serial data adapters (SSDAs) such as the 6852 require special treatment. The testability guidelines for this device, and for devices like it, include significant software initialization. The initialization sequence requires access to control and data lines and can be illustrated as follows:

RESET* (drive low then high)
Load control register 1
Load control register 2
Load control register 3
Load sync control register

Transmit and receive logic should be inhibited via bits 0 and 1 in control register 1 before the contents of any other control register (or the sync register) are modified.

Other 6800 microprocessor family devices, such as the 6828 priority interrupt controller, the 6840 programmable counter timer, and the 6801 microcomputer, are also used extensively. These devices are not covered here due to their similarity to support devices of other families covered elsewhere.

THE 2901 MICROPROCESSOR FAMILY

The 2901 is a bit-slice microprocessor made with bipolar technology, as opposed to the metal oxide semiconductor (MOS) technology used to fabricate the other microprocessors and microcomputers described earlier. As such, it is not, in and of itself, a microcomputer. Nor is it dynamic in nature, because it can run asynchronously. The 2901-based printed circuit boards do not normally require refresh circuitry (with its attendant testability problems), unless dynamic RAM is used elsewhere in a system design. Each 2901 package features a 4-bit data input, 4-bit data output (which is tristateable), RAM address and shift logic inputs, plus several status and control lines.

The key guideline to testability of a board using 2901 devices is individual control of the OE* lines. With control, each device can first be individually exercised before the full functional test of the board. Access should also be provided to the following common lines:

D0–D3	Data input lines
Y0–Y3	Data output lines
A0–A3	Local RAM A address lines
B0–B3	Local RAM B address lines
RAM–RAM3	RAM shift logic
Q0–Q3	Q register shift logic
CH	Carry-in
CP	Clock pulse input
F	Zero status

The Y0–Y3 lines are tristate lines and should have pull-up resistors attached to the common bus. This same guideline holds for the F output, which is an open collector output.

The ability to control individual OE* lines cannot be overemphasized from a testing standpoint. This individual control allows a partitioned testing approach and reduces the need for using 2901 microcode for a complete board test. This control also reduces the effort required for software initialization of the A and B latches and the Q registers.

The 2903 has the same data input port as the 2901 (DA0–DA3), but it has two bidirectional output data ports (called DB0–DB3 and Y0–Y3). Testability guidelines for the 2903 include those mentioned for the 2901, plus the need for control and access points at EA*, WE*, OEB*, OEY*, IEN*, and LSS*

The 2902 carry lookahead device creates parallel carry inputs for slices beyond the least significant slice. This device will normally be tested as part of the 2901/2903 system design and cannot be isolated. Test point access is helpful from a diagnostic standpoint, but, since inputs to the 2902 do not come from tristate outputs of the 2901/2903, control points serve no useful function.

The 2910 microprogram sequencer primarily requires control point access to the CCEN* (condition code enable) and OE* lines. The clock pulse (CP) is synchronous with other 29XX devices on the UUT.

The 2930 and 2932 program control units are architecturally similar to the 2901, with Q register functions modified to operate as a local data register and the local RAM used as a stack internal to the devices.

The 2930 is the preferred device since it provides for partitioning via the OE* signal. The 2932 also lacks the flexibility of the IEN* (instruction enable) and RE* (register input enable) functions, which are desirable from both system (board) functional design and testability standpoints.

Testing of 2901 series boards is not difficult as long as access to control points (primarily OE* lines) is provided in PCB design.

THE 68000 PROCESSOR FAMILY

The 68000 processor, although physically very large, is much easier to test than some of the other processors because it has a larger fan-in and fan-out (i.e., more I/O pins). When applied at the circuit board level, the following control and visibility points should be made accessible.

Control Points
- CLK input if normal speed exceeds ATE speed
- DTACK* and VPA so that the ATE can handshake with the 68000
- RESET* so that the processor can be reset under external control
- HALT* so that the processor can be stopped without having to perform a reset and initialization sequence
- BR* and BGACK* so that the external ATE can become the bus master; BERR* so that the error function can be checked

Visibility Points
- Status lines FC0–FC2
- The E line to 6800 family devices, since it runs asynchronous to the bus and must be synched up to by the ATE
- BG* so that the ATE will know that it is about to be granted the bus
- AS*, LDS*, UDS*, R/W*, and VMA* so that the ATE can strobe at the proper time and decode reads, writes, address, and data

The 68008 Processor

The 68008 is an 8-bit version of the 68000, and there are fewer signals needed to decode address and data information. The other control and visibility signals remain the same as for the 68000. Control points include BR*, VPA*, BERR*, DTACK*, RESET*, and HALT*. Visibility

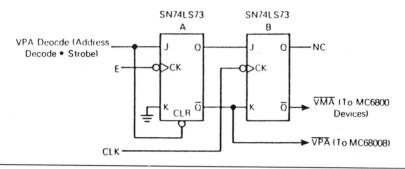

FIGURE 6-27.   External VMA generation. Extra circuitry is often required on-board to decode signals that are not specifically provided as dedicated signal pins on a processor. Any signals required for this type of function are valuable visibility points.

points include the status lines FC0–FC2, the E (enable) line, and the read/write, address, and data strobe lines R/W*, AS*, and DS*.

Access to the address and data busses is a firm testability requirement as with any LSI/VLSI device. Since the 68000 and 68008 are asynchronous rather than synchronous devices, using them with 6800 family peripheral devices requires a synchronization line. With the 68000, that line is VMA*. There is no VMA* line for the 68008, so it must be created. The equivalent VMA* signal is decoded using the clock, the E line, and VPA decode, which is an active high decode indicating that AS* has been asserted and the address bus is addressing a 6800 peripheral. The VPA* output is fed back to the 68008 to indicate that the data transfer should be synchronized with the E signal (see Figure 6-27). This type of decode and signal generation is typical of that required in the ATE interface in order to perform testing, and illustrates clearly the need for the visibility and control points identified for the 68008.

The 68451 Memory Management Unit

Some MMU lines are inputs and outputs, meaning that the bus buffers can be directed inward to direct information to the MMU or outward to drive the bus. Some are wire-ORed, and some are wire-ANDed. The IRQ*, FAULT*, and AN* lines are active low and wire-ORed in a multiple MMU system.

The ALL pin is an active high pin (open drain) and will be wire-ANDed in a multiple MMU system. It must have a pull-up resistor. The GO* and MAS* signals are not open drain but can be tristated, so they also need pull-ups to hold the signals inactive in the high-impedance state. Critical visibility points include ANY* and ALL. Critical control points include CS*, the chip select line, RESET*, and the MODE input, which controls the function of the MAS* signal. It is also helpful to have control of, or at least visibility to, the ED*, HAD*, IRQ*, and FAULT* lines.

The 68440 Dual Direct Memory Access Controller

The dual DMA controller (DDMA) is designed to move blocks of data with a minimum amount of intervention by the host processor. Any time a device is going to do something with "minimum intervention" by an external device, it needs to be controlled for testability purposes.

In addition to the CS* line, a standard control requirement, control is needed of the BEC0*–BEC2* bus exception control lines to tell the DDMA to reset (there is no hardware reset line), to halt (there is no halt line), or to retry. Control is also needed for the DONE* line, to tell it to terminate an operation, and the REQx* and PCLx* lines. REQ* inputs allow the ATE to request a DMA operation, while PCL* lines can serve the hold function for input from a slow device (e.g., an ATE system).

Visibility is required to DBEN*, DDIR*, DTC*, and the ACKx* lines.

One of the most difficult fault isolation tasks is to determine where in a bus-oriented system a defect actually resides. Having control of the bus drivers and transceivers allows immediate isolation at least to a section of the circuitry. Extra gates can be added to the individual input lines to provide this control.

The 68450 Direct Memory Access Controller

This device is similar in function to the 68440 DDMA and the testability requirements are virtually identical. Control is needed of CS*, DONE*, and each of the four REQx* and PCLx* inputs, as well as the BEC0*–BEC2* lines. Visibility is needed to the DBEN* and DDIR* lines, the DTC* (device transfer complete), and the four ACKx* output lines. Separate control of bus drivers and transceivers is important when semiautonomous devices such as these are used on a printed circuit board.

The 62661/68661 EPCI

The 62661/68661 is an enhanced programmable communications interface (EPCI). It supports bisynchronous and asynchronous operations. Whenever the word *asynchronous* appears, it should be a flag that there is a possible testability problem lurking in the wings. The EPCI does, however, include loopback, both local and remote, to make testing a little easier—if control is provided.

Control points needed for the EPCI include RESET and CE* as well as BRCLK if the internal baud rate generator is used. In addition, it is important to be able to control CTS*, DCD*, and DSR*. DCD* and CTS* are needed to separately operate the receiver (DCD*) and transmitter (CTS*). Required visibility points include TxC*/SYNC, RxC*/BKDET, TxEMT/DSCHG* as well as the ready outputs TxRDY* and RxRDY*. Control points for any device are inputs which control reset, chip select, and clock functions as well as other enables (e.g., DSR*, DCD* and CTS*). Visibility points are status outputs (TxRDY* and RxRDY*) as well as clock outputs and synchronization lines.

Testing a device like this at board level involves having enough control and visibility to check each internal block and to be able to disconnect (electrically) the device from the board when it is not to be used (e.g., while other devices are being tested).

The 68230 PI/T

The 68230 parallel interface/timer (PI/T) provides uni- and bidirectional 8-bit and 16-bit parallel interface capability, can be DMA driven, and includes its own timing generator for elapsed time measurement and watchdog timer functions.

The I/O ports and the timer are logically independent (read asynchronous!) on the chip. The on-chip square-wave generator puts out a signal that must be controllable via the CLK input, so CLK is a required control point. The other required control points are the RESET* and CS lines and H1, the handshake input line.

Visibility points include the other handshake lines, although H3 could be identified as control and H2 and H4, being bidirectional, are candidates for control as well.

The block diagram for the 68230 PI/T (Figure 6-28) illustrates the reason why the various control and visibility lines have been chosen. Notice also that the lines that have been selected are the critical lines only; there is a lot of opportunity for providing control of additional

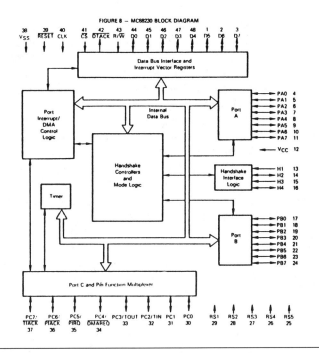

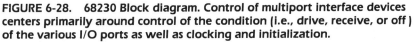

FIGURE 6-28.   68230 Block diagram. Control of multiport interface devices centers primarily around control of the condition (i.e., drive, receive, or off) of the various I/O ports as well as clocking and initialization.

lines as well. Additional lines which would be nice to have control of and visibility to include all of the pins across the bottom of the block diagram (device pins 25 through 37). Any time there are dual-function (e.g., input/output) lines such as pins 30 though 37, control should be considered.

The 68681 DUART

The 68681 is a dual universal asynchronous receiver/transmitter (DUART)—an I/O device that can be used with a variety of on- and off-board peripheral devices. The critical control points are the standard RESET* and CS* inputs, and the critical visibility point is DTACK*, which tells the system that a device transfer is complete.

As with any communications-type device, it is also helpful to have control of, or at least synchronization to, the clock input. And in a normal configuration where the TxDA, TxDB, RxDA, and RxDA lines come off the board to interface with external devices, control, and visibility would be present via the normal edge connector or other interface.

Should the 68681 be used to communicate with other on-board devices, some means of controlling the RxDA and RxDB lines should be provided, and visibility will be required to the TxDA and TxDB lines. Other helpful lines, although they are not strictly necessary, are the "local" reset lines RS1 through RS4.

Figure 6-29 is a modified reproduction of a diagram from the manufacturer's data sheet that shows how the 68681 is normally connected to a 68000 processor. The modifications made to make it testable are the two two-input AND gates marked with asterisks (*). Without the extra control input, it is very difficult to perform a partitioned test on a board containing this device.

In fact many of the suggested design examples in device manufacturers' data sheets are typically untestable. Great care must be taken

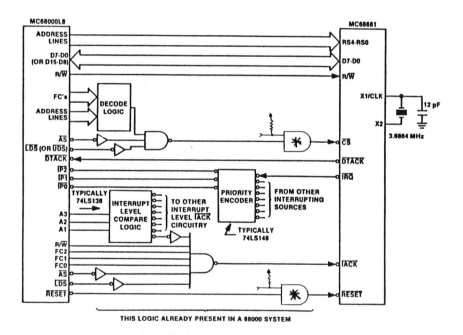

THIS LOGIC ALREADY PRESENT IN A 68000 SYSTEM

FIGURE 6-29. MC68000 to MC68681 interface. Device manufacturers data sheets and application notes often show difficult to test implementations. The gates marked with an asterisk (*) have been changed from one-input gates to two-input gates to provide separate control of the MC68681.

when "borrowing" these designs (instead of reinventing them) to make sure that they are modified to be testable. When one imagines the millions of copies of untestable circuit configurations printed by the world's device manufacturers, one ceases to wonder why so many untestable configurations continue to appear in new designs. These configurations are, after all, printed by reputable companies with good products. And design engineers do not typically reinvent the mundane parts of a circuit; they concentrate instead on the value-added functions for their specific application. Thus the readily available (untestable!) circuits end up being multiplied.

The 68652 MPCC

The 68652 multiprotocol communications controller (MPCC) is an even more sophisticated device than the DUART in that it provides for internal formatting between transmission and reception of data. What this means from a testability standpoint is that the data coming out may not resemble the data going in!

In addition to the standard RESET and CE control inputs, we would like control of the DBEN line so that we can tristate the DB00–DB15 lines going to the 68000 processor. And, in addition to the clocks RxC and TxC, it is very helpful to have access to and control of the maintenance mode (MM) line. With the MM line, we can loop TxSO back to RxSI and TxC* into RxC for off-line (i.e., disconnected from the other I/O devices) diagnostic purposes.

The main visibility point required is S/F, the SYNC/FLAG line. This line is asserted for one RxC clock time whenever a SYNC or FLAG character is detected.

The 68653 Polynomial Generator/Checker

Another useful device in the communications environment is the 68653 polynomial generator/checker (PGC). The 68653 PGC is used to generate and check bit block and cyclic redundancy codes to detect, and allow for correction of, bit errors during communications using a device like the MPCC.

An analysis of the block diagram for the device (Figure 6-30) reveals the need for control of at least one of the CHIP ENABLE lines (CE0* or CE1*). Control of A0 and A1 is also necessary, but this is usually available via the processor which we previously made controllable. Notice that there is no hardware reset line for this device, so software

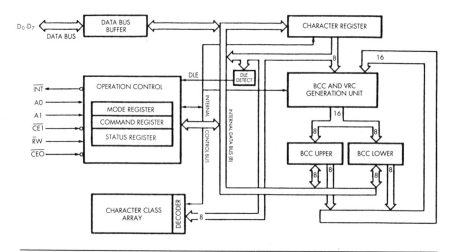

FIGURE 6-30.   68653 PGC block diagram. Devices that generate (and/or compare) polynomials should be able to be separately deselected (via their chip select or address inputs) in order to partition the board they are designed into.

initialization is required to set all of the internal registers into known states. The primary visibility point needed is INT*, since we wish to be able to prevent it from getting to some other device on the board which we may be in the process of testing.

The 68120 IPC

Testability requirements include the control points CS*, RESET*, CLK, and TIMER IN and the visibility points DTACK* and TIMER OUT.

The 68454 IMDC

Testability requirements include the control points BG*, CS*, and RES* and the visibility points DTACK*, EN0, EN1, LOCAL, OWNX, and DDIR.

The 68465 Floppy Disk Controller (FDC)

Testability requirements include the control points CLK, RESET*, CS*, ACK, and DONE* and the visibility points DTACK* and REQ*.

The 68561 MPCC-II

Testability requirements include the control points CS*, RESET*, and EXTAL and the visibility points DTACK* and BCLK.

The 68562 DUSCC

Testability requirements include the control points RESET*, CS*, X1/CLK, and X2/IDC* and the visibility points DTACK*.

The 68590 LANCE

Testability requirements include the control points HLDA*, READY*, RESET*, CE*, and RS and the visibility points ALE/AS* and HOLD/BUSRQ*.

The 68901 Multifunction Peripheral (MFP)

Testability requirements include the control points CS*, RESET*, IEI*, SI, SO, RC, and TC and the visibility points DTACK*, IRQ*, and the timer outputs.

THE 68020 PROCESSOR

The 68020 is the first full 32-bit implementation of the 68000 family. Previous implementations (i.e., 68000, 68010) could handle 32-bit data, but used multiplexed 16-bit internal busses to do so. The 68020 has an asynchronous bus structure and supports a dynamic bus sizing mechanism that allows the processor to transfer operands to or from external devices while automatically determining device port size on a cycle-by-cycle basis. The functional signal groups for the 68020 are illustrated in Figure 6-31.

The control and visibility points for this device are similar to those for the earlier 68000 version. The control lines needed for the 68020 include

- RESET*, in order to initialize the processor
- HALT*, in order to provide for single-stepping on the external bus lines

- BR*, in order to let an external device become a bus master
- BGACK*, in order to let the processor know that the external device has become a bus master
- IPL2*–IPL0*, the interrupt signals
- DSACK0*, DSACK1*, and BERR*, in order to control bus cycle termination
- CDIS*, to disable on-chip cache to assist emulator support

Visibility points needed include

- FC0–FC2, to identify the address space of each bus cycle
- SIZ0–SIZ1, to indicate the number of bytes remaining to be transferred for the current cycle
- R/W*, to tell whether reading or writing
- AS*, to tell when data are valid on the address bus
- DS*, to tell when data are valid on the data bus

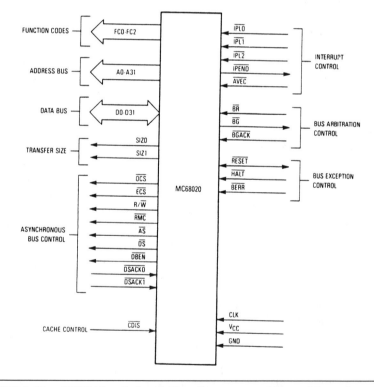

FIGURE 6-31.   68020 signal description. Processors with on-chip cache functions normally have one (or more) pins that can be used to control those functions. In the case of the 68020, that pin is the CDIS* pin.

- BG*, to grant an external device bus mastership
- ECS*, to indicate that an external cycle is starting
- OCS*, to indicate that an operand cycle is starting
- DBEN*, provides an enable signal to external buffers

THE 68030 PROCESSOR

The 68030 is a second-generation full 32-bit microprocessor from Motorola. It combines a CPU core with a data cache, an instruction cache, an enhanced bus controller, and a memory management unit, all on the same piece of silicon. A summary of its functional signals is illustrated in Figure 6-32.

Since this device has so many functions on-chip, there are additional requirements for controlling them. Rather than repeat the guidelines for the 68020, the following guidelines point out the additional

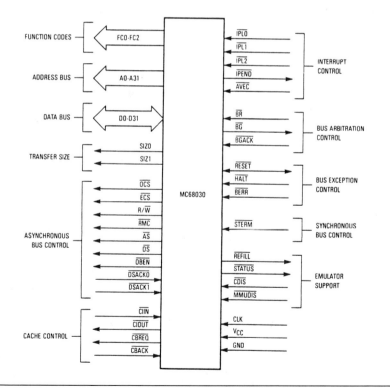

FIGURE 6-32.   68030 signal description. In addition to initialization and bus control, control is needed of the multiple cache input pins on a device like the 68030.

signals required for good partitioning, control, and visibility of the chip when it is installed on a board.

Additional Control Signals
- STERM*, indicates that 32 bits of data may be latched on the next falling clock edge
- MMUDIS*, disables the MMU cache to support emulation
- CIIN*, prevents data from being loaded into the data and instruction caches of the 68030

Additional Visibility Points
- STATUS*, indicates the state of the microsequencer
- REFILL*, indicates when the processor is beginning to fill the pipeline

The MMUDIS*, CIIN*, STATUS*, and REFILL* signals, while designed to support design verification and debug using emulation equipment, can also be used to support test and troubleshooting because an ATE can be programmed to operate like an emulator. Thus these signals should be made physically or electrically accessible if test is to take advantage of them. The input signals should not be tied hard to V_{cc} when not needed functionally (i.e., after the design is complete!). The output signals should similarly not be left unconnected (and therefore not visible for test purposes).

THE 88000 RISC PROCESSOR FAMILY

The latest microprocessor technology to be introduced is called *reduced instruction set computer* (RISC) technology. These processors are stripped-down streamlined versions of advanced processors and are designed to execute fewer instructions than most complete instruction set computers (CISCs), but to execute those instructions at a very high throughput rate.

The guidelines for RISC processors and their peripheral circuits are, however, no different from those for any other processor (as illustrated with the 88100 and 88200 in Figures 6-33 and 6-34, respectively). Resets, clocks, and enables for control and handshake and status outputs for visibility are the order of the day.

THE 320C2x DSP DEVICE FAMILY

Another very popular family of devices is the digital signal processing (DSP) chip family. These chips combine the flexibility of a high-speed

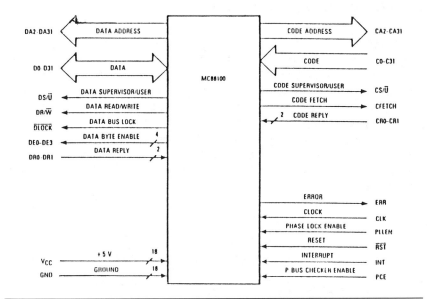

FIGURE 6-33. RISC processor block diagram. RISC processors, although they operate differently from their predecessors, have similar control and visibility requirements when it comes to testability. The P BUS CHECKER and PHASE LOCK ENABLE lines are important control points.

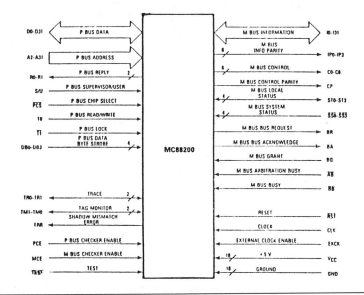

FIGURE 6-34. RISC processor companion chip block diagram. Companion, or support, devices should also be separately controllable. In the case of the 88200, the CHECKER ENABLE lines (P Bus and M Bus) and the TEST line need control in addition to the regular control lines.

microcontroller with the numerical capability of an array processor. The 320C2x family of devices from TI includes on-chip CPU, RAM, ROM, instruction cache, serial ports, and timers. These programmable devices can often be used in place of custom VLSI devices and bit-slice processors for signal processing applications. A block diagram of the generic family of TI's DSP chips is shown in Figure 6-35.

The controller portions of a DSP chip resemble the standard functions and lines of a typical processor device. There are additional functions on-chip, however, which require their own control and visibility considerations.

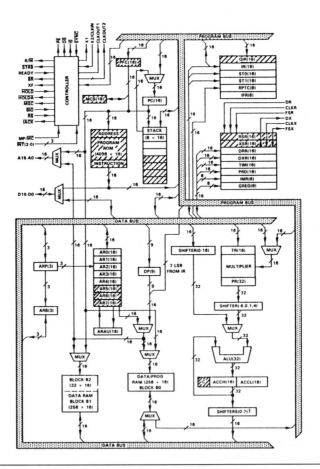

FIGURE 6-35. TMS320C2x block diagram. DSP chips have both controller and array processor capabilities. Control lines for the controller portion are similar to the control lines for any microprocessor because all interfacing with the internal array processor is done through the controller.

In the area of standard control points, these devices would need to be controllable at their reset (RS*) and hold inputs (HOLD*). In the area of special concerns, the MP/MC* pin on the 320C25 (only) has the function of an "external access" pin on an 8048 family device. If MP/MC* is in the logic 1 state, external ROM is used to run the device. If it is held low, internal ROM is used. For the 320C20, this is a moot point—the MP/MC* line must be tied high.

Additional handy control points for this device include the clock input (X2CLKIN); the READY line, which facilitates handshaking be tween the 320C2x and other devices (including ATE!), and the SYNC* input, which allows clock synchronization of two or more TMS320C2x devices.

Visibility will be required for the read/write line (R/W*), the strobe line (STRB*), bus request (BR*), the clock outputs (CLKOUT1 and CLKOUT2), and the MSC* line, which indicates that the device has just completed a memory operation (such as an instruction fetch or a data memory read/write). The MSC* line can be used to generate a one-wait-state READY signal for slow memory.

While testing these devices is certainly a challenge, testing boards that contain them need not be a tremendously difficult task as long as the devices themselves, and the devices they are connected to, are designed with partitioning in mind and have the proper control and visibility points accessible.

MERCHANT SEMICONDUCTOR USE GUIDELINES SUMMARY

No matter what the function or complexity of the device, when it is installed on a board it must be controllable (for partitioning of the design verification, fault simulation, and testing tasks) and its critical outputs must be observable in some way by whatever test resource (whether ATE or BIT) is employed to test it in the factory and in the field.

The purpose of the preceding walk through many devices was to familiarize each designer with the *types of signals needed for control and visibility* on any type of merchant LSI/VLSI device. It is virtually impossible to document all of the guidelines for all of the available devices and those likely to be available. But just as boards are tending to look alike today, devices will tend to look alike tomorrow.

New devices made from megacells will result in systems integrated into single chips, and those new devices will be installed on boards to create complex, powerful systems. Maintaining the critical testability attributes (partitioning, control, and visibility) at board and system levels will become even more important.

Each designer, in conjunction with his or her test engineer, should consult the data sheets for new devices to be certain that the critical control and visibility points for those devices have been properly identified and are easily available for simulation and test purposes when those devices are designed into board and systems.

7

LSI/VLSI ASIC
Level Techniques

Today, with the extensive utilization of LSI and VLSI technology, and with the explosive growth in the availability of semicustom and full-custom customer-specific ICs (CSICs) for virtually any application, it has become apparent that even more care will have to be taken during the component design stage in order to insure adequate testability and producibility of digital ICs themselves and of the networks they are assembled into. This proliferation of complex devices with many varied functions has led to the need for rigorous and highly structured device design practices if adequate tests are to be generated in a timely and cost-effective manner.

Many "silicon foundries" and application specific IC manufacturers have actually designed software tools into their design environments that will take an initial (and typically untestable) unstructured design and synthesize either a complete, or sometimes a reduced intrusion, scan path into it. The software tools replace general memory elements (e.g., flip-flops, shift registers, and the like) with so-called scannable versions of these devices and connect the devices in series to form a *scan chain* that is used for partitioning, controlling, and observing the device under test.

Most structured design practices are built on the concept that if the values of all of the latches internal to the design can be controlled to specific values and observed with a straightforward operation, then the task of test generation, and possibly fault simulation, can be reduced to that of doing test generation and fault simulation for a combinational logic network. A control signal can switch the memory elements from their normal mode of operation to a mode that makes them controllable and observable.

This chapter discusses the most widely used of the structured, or scan design, methods at the device level and concludes with a discussion of how the scan attributes of the various devices can be used at the board level to simplify all phases of the testing process. Each method can provide significant testability advantages compared to unstructured device designs.

LEVEL-SENSITIVE SCAN DESIGN (LSSD)

Level-sensitive scan design (LSSD) is IBM's discipline for structural design for testability. Scan refers to the ability to shift any state into or out of the network. Level sensitive refers to constraints on circuit excitation, logic depth, and the handling of clocked circuitry. A key element in the design is the shift register latch pair (SRL), which can be implemented as illustrated in Figure 7-1. Such a circuit is immune to most anomalies in the AC characteristics of the clock, requiring only that it remain active at least long enough to stabilize the feedback loop before

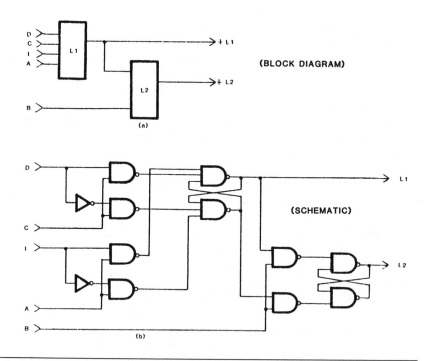

FIGURE 7-1.   Level-sensitive scan design elements. The extra inputs (I, A, and B) to the shift register latch pair allow each latch to be set directly. The state of the latch can be observed at the L2 output.

being returned to the inactive state. Lines D and C form the normal mode memory function, while lines I, A, B, and L2 compose additional circuitry for the shift register function.

The basic idea behind using the extra circuitry, which can consume up to 20 percent of the available functional circuitry on an IC, is to make sequential circuitry look combinatorial—that is, to be able to force states at any node without "clocking" and to similarly be able to observe the state of any node. The extra inputs allow any latch to be set or reset by cycling the A and B clock inputs between the logic 0 and logic 1 levels (i.e., without clocking). While the wording above may seem strange at times, the design of the circuitry allows automatic test generation programs to treat even fully sequential circuitry as simple combinatorial circuitry. LSSD designs essentially fool the software algorithms, by making sequential circuitry even more sequential, into thinking that the circuitry is combinatorial.

Figure 7-2 shows the familiar generalized sequential circuit model modified to use the LSSD shift register latches. This technique provides both controllability and observability, allowing the testing to be augmented by controlling inputs and internal states and easily examining the resulting internal state behavior. A disadvantage is the serialization of the test, which could potentially take more time for test execution. An advantage is that only four extra physical device pins are required for the testability interface. On very large ICs, however, multiple scan chains are often implemented to partition the device and reduce the length of the test vectors. These multiple scan chains may use either

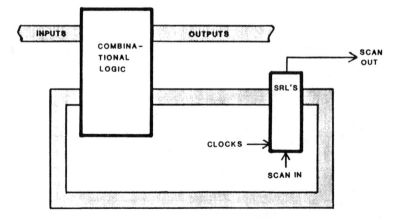

FIGURE 7-2.   Generalized LSSD approach block diagram. An LSSD design can be thought of as a collection of combinational logic interconnected to the SRL block. Thus the SRLs act as additional pseudo-primary inputs and outputs.

extra dedicated physical device pins or on-chip multiplexing under the control of additional scan registers for selection of the desired scan chain.

The shift registers that make up each scan chain are cascaded by connecting each L2 line to the following I line, and are operated by clocking lines A and B in two-phase fashion. Figure 7-3 shows three SRLs connected in series. If it were desired, for example, to force the middle circuit into the logic 1 state, a logic 1 would be placed on the SCAN IN input and the A and B clocks would have their levels cycled from logic 1 to 0 and back again twice. If it were desired to ascertain the logic state of the middle latch at any time, the A and B lines would be cycled to place the middle latch's +L2 state on the SCAN OUT line.

Also shown in Figure 7-3 is the example of four modules (or chips) cascaded for shift register action. Note that each IC could include an SRL chain and, one level up, a board could contain cascaded SRL-equipped ICs. Each level of packaging requires only the same four additional lines to implement the shift register scan feature.

Figure 7-4 depicts a general structure for an LSSD subsystem with a

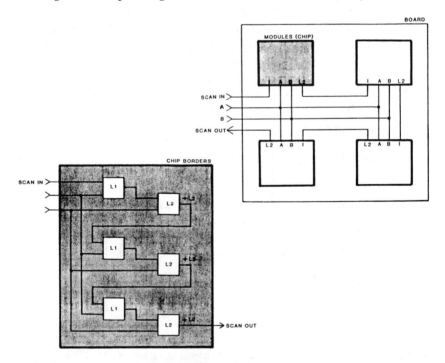

FIGURE 7-3.   Cascaded LSSD registers and modules. Individual latches can be cascaded to form an on-chip scan chain. At the board level, individual chips can be connected in series to form a board level scan bus.

two-phase system clock. It is not practical to implement RAM with SRL memory, so additional procedures are required to handle embedded RAM circuitry. In extremely complex combinatorial networks, it is both possible and advisable to add SRLs to specific nodes within them as well (e.g., fan-in and fan-out points).

The LSSD design philosophy has some negative impacts on cost and performance. First of all, the shift register latches in the shift register are two or three times as complex as simple latches. Up to four additional primary inputs/outputs are required at each package level for control of the shift registers. External asynchronous input signals must not change more than once every clock cycle. Finally, all timing within the subsystem is controlled by externally generated clock signals.

The LSSD structured design approach for design for testability can, however, alleviate some of the problems in designing, manufacturing, and maintaining LSI/VLSI systems at a reasonable cost and should be considered as one of the available options.

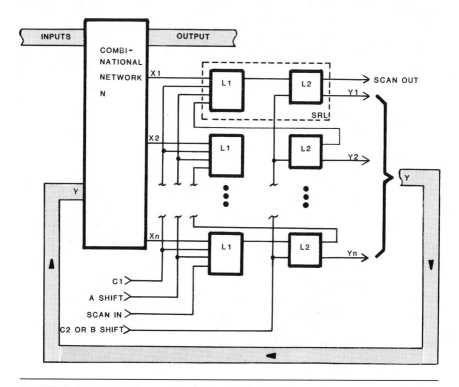

FIGURE 7-4.   General structure for an LSSD subsystem. This diagram shows how the outputs of the combinational network are interconnected to the scan registers whose outputs, in turn, are fed back to the inputs of the combinational network.

SCAN PATH

The *scan path* technique has the same objectives as the LSSD approach but uses a different circuit design for the memory elements. The memory elements used in the scan path approach are shown in Figure 7-5. The memory element is called a raceless *D*-type flip-flop with scan path.

In system operation, clock 2 is at a logic 1 level for the entire time. This, in essence, blocks the test or scan input from affecting the values in the first latch. This *D*-type flip-flop really contains two latches. Also, by having clock 2 at a logic 1 level, the values in latch 2 are not disturbed.

Clock 1 is the sole clock in system operation for the *D*-type flip-flop. When clock 1 is at the logic 0 level, the system data input can be loaded into latch 1. As long as clock 1 is in the logic 0 state for sufficient time to latch up the data, it can then be returned to the logic 1 state. As it turns off, it then will make latch 2 sensitive to the data output of latch 1. As long as clock 1 is at a logic 1 long enough so that data can be latched into latch 2, reliable operation will occur.

In terms of the scanning function, the *D*-type flip-flop with scan path has its own scan input called the test input. This is clocked into latch 1 by clock 2 when clock 2 is in the logic 0 state, and the results of

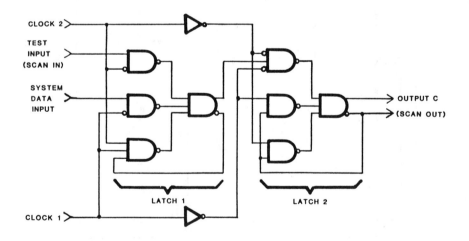

FIGURE 7-5.   Scan path approach circuit implementation. The latch configuration for the scan path circuit differs from that of an LSSD circuit, but its function is very similar.

latch 1 are clocked into latch 2 when clock 2 is a logic 1. Again, this applies to master/slave operation of latch 1 and latch 2 with its associated race. With proper attention to delays, this race will not be a problem.

Another feature of the scan path approach is the configuration used at the logic card level. ICs on the logic card are all connected into a serial scan path such that for each board there is one scan path. In addition, there are gates for selecting a particular card in a subsystem. In Figure 7-6, when both the X and Y inputs are in the logic 1 state, clock 2 will be allowed to shift data through the scan path. At any other time, clock 2 will be blocked and its output will be blocked. The reason for blocking the output is that a number of card outputs can then be connected together; thus, the blocking function will put all unselected card outputs to noncontrolling values so that a particular card can have unique control of the unique test output for that system.

Other than the lack of the level-sensitive attribute to the scan path approach, this technique is very similar to the LSSD approach. The scan path approach was the first practical implementation of the shift register technique for the testing of complete systems.

SCAN/SET LOGIC

A technique similar to scan path and LSSD, but not exactly the same, is the scan/set technique. The basic concept of this technique is to use shift

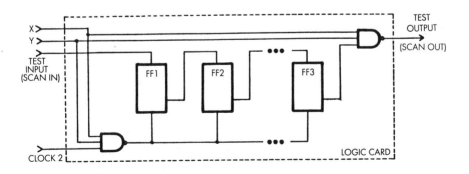

FIGURE 7-6.   Board level scan path implementation. Scan path latches can be configured at the board level. The figure also shows gates for implementing an addressing scheme for the board's scan path.

registers, as in scan path or in LSSD, but these shift registers are not in the data path. That is, they are not in the system data path; they are independent of all the system latches. Figure 7-7 shows an example of the scan/set logic, sometimes referred to as *bit serial logic*.

The basic concept is that the functional network can be sampled at up to 64 points. These points can be loaded into the 64-bit shift register with a pulse on the LOAD line. Once the 64 bits are loaded, a shifting process will occur, and the data will be scanned out through the scan-out pin in response to inputs to the CLOCK line. For the set function, the 64 bits can be funneled into the system logic, and then the appropriate clocking structure required to load data into the system latches is required in the system logic itself. Furthermore, the set function could also be used to control different paths to ease the testing function. In essence, 64 control and/or visibility points can be gained at the price of as few as four interface pins plus the shift register.

In general, this serial scan/set logic would be integrated onto the same chip that contains sequential system logic. Some applications, however, have been put forth where the bit serial scan/set logic was off-chip, and the bit-serial scan/set logic only sampled outputs or drove inputs to facilitate in-circuit testing.

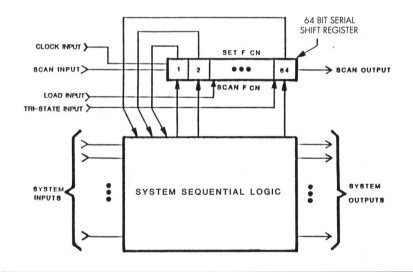

FIGURE 7-7.   Scan/set logic implementation concept. With scan/set logic, the shift register is not an integral part of the functional circuitry. It "sits beside" the functional circuitry, but can control and observe it.

It is not required that the set function set all system latches or that the scan function scan all system latches. This design flexibility does, however, have an impact on the software required to implement and support such a technique. In many ways scan/set logic implementation is similar to the *incomplete scan* and *reduced intrusion scan path* techniques. While it does provide for partitioning, control, and visibility, it does not provide the absolute structure that makes test pattern generation, using today's tools, fully automatic.

Another advantage of this technique is that the scan function can occur during system operation. That is, the sampling pulse to the 64-bit serial shift register can occur while system clocks are being applied to the system sequential logic so that a snapshot of the sequential machine can be obtained and off-loaded without any degradation in system performance. On-line monitoring of circuit functions can thus be implemented with variations on the scan/set logic theme.

RANDOM ACCESS SCAN

Another technique similar to the scan path technique and LSSD is the *random access scan* technique. This technique has the same objective as scan path and LSSD—that is, to have complete controllability and observability of all internal latches. Random access scan differs from the other two techniques in that shift registers are not employed. What is employed is an addressing scheme that allows each latch to be uniquely selected so that it can be either controlled or observed. The mechanism for addressing is very similar to that of a random access memory.

Figures 7-8 and 7-9 show the two basic latch configurations that are required for the random access scan approach. Figure 7-8 shows a single latch which has added to it an extra data port, which is the SCAN DATA IN (SDI) port. These data are clocked into the latch by the SCK clock. The SCK clock can only affect this latch if both the X and Y addresses are at the logic 1 state. If they are, then the SCAN DATA OUT (SDO) port can be observed.

System data, labeled DATA in Figures 7-8 and 7-9, are loaded into this latch by the system clock labeled CK.

The set/reset-type addressable latch in Figure 7-9 does not have a scan clock to load data into the system latch. This latch is first cleared by the CL line, and the CL line is connected to other latches that are also set/reset-type addressable latches. This then makes the output value Q

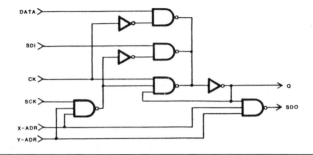

FIGURE 7-8. Random access scan latch configuration 1. An address is implemented for each and every latch with the random access scan approach. This diagram shows the normal D-type latch.

a 0. If latches are required to be set to a 1 for a particular test, a preset is directed at those latches. This preset is directed by addressing each one of the appropriate latches and applying the preset pulse, labeled PR. The output of the latch Q will then go to a 1.

The observability mechanism for SCAN DATA OUT is exactly the same as for the latch shown in Figure 7-8.

Figure 7-10 gives an overall view of the system configuration of the random access scan approach. Notice that basically there is a Y address, an X address, a decoder, the addressable storage elements, which are the memory elements or latches, and the sequential machine, system clocks, and CLEAR function. There is also a SCAN DATA IN, which is the input for a given latch, SCAN DATA OUT, which is the output data for that given latch, and a scan clock. There is also one logic gate to create the preset function.

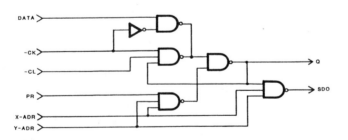

FIGURE 7-9. Random access scan latch configuration 2. A variation of the D-type latch which includes preset and clear functions.

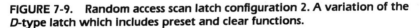

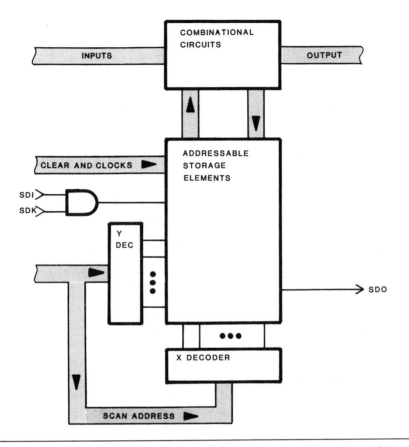

FIGURE 7-10.   Overall random access scan configuration. Illustration of how an overall circuit incorporating random access scan is interconnected. Extra inputs are required for the scan addresses.

The random access scan technique allows the observability and controllability of all system latches. In addition, any point in the combinational network can be observed with the addition of one gate per observation point plus one address in the address gate per observation point.

BUILT-IN LOGIC BLOCK OBSERVATION (BILBO)

BILBO takes the scan path and LSSD concepts and integrates them with the signature analysis concept. The BILBO registers are used in the system operation as shown in Figure 7-11. Basically, a BILBO register with combinational logic, as well as the output of the second combinational logic, can be tested with reasonable thoroughness with pseudorandom patterns. Thus, if the inputs to the BILBO register can be controlled to fixed values so that the BILBO register is in the maximal length, linear feedback, shift register mode (signature analysis), it will output a sequence of patterns which are very close to random patterns. Thus, random but repeatable patterns can be generated quite readily from this register. These sequences are called *pseudorandom number* (PN) *patterns*.

If, in the first operation, the BILBO register on the left in Figure 7-11 is used as the PN generator, then the output of the BILBO register will be random patterns. This will then do a reasonable test, if sufficient numbers of patterns are applied, of the combinational logic network 1. The results of this test can be stored in a signature analysis register with multiple inputs to the BILBO register on the right. After a fixed number of patterns have been applied, the signature is scanned out of the BILBO register on the right. This register will now be used as a PN sequence generator. The BILBO register on the left will then be used as a signature analysis register with multiple inputs from combinational logic network 2, as illustrated in Figure 7-12.

In this mode, the combinational logic network 2 will have random patterns applied to its inputs and its outputs stored in the BILBO register on the far left. Thus, the testing of combinational logic networks 1 and 2 can be completed at very high speeds by applying only the shift clocks while the two BILBO registers are in the signature analysis mode. At the

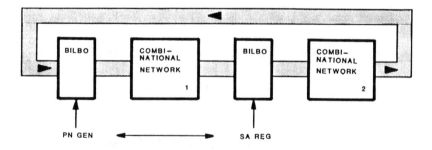

FIGURE 7-11.   BILBO testing of network 1. The BILBO register on the left is configured as a pattern generator while the one on the right becomes a signature analyzer.

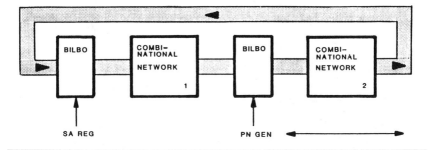

FIGURE 7-12.   BILBO testing of network 2. The BILBO register on the left is configured as a signature analyzer, while the one on the right is set as a pattern generator.

conclusion of the tests, off-loading of patterns can occur and determination of good machine operation can be made.

This technique solves the problem of test generation and fault simulation if the combinational networks are testable via pseudo-random patterns. There are, however, some known networks which do not fall into this category. These are primarily programmable logic arrays (PLAs) as shown in Figure 7-13. The reason for this is that the fan-in of PLAs is too large. Random combinational logic networks with a maximum fan-in of four can do quite well with pseudorandom patterns. Any of the output scan techniques, however, can be used with networks like PLAs.

BILBO should thus be applied selectively to the logic circuits most suited to its use. Other scan methods should then be applied as well to those circuits not capable of being economically or efficiently tested using the BILBO structure.

SIGNATURE ANALYSIS

Signature analysis is often used as an adjunct to several testing methods, including in-circuit and functional, and serves basically as a data compression technique. Data compression is achieved in the signature analyzer by probing a logic test node from which data are input for each circuit clock cycle that occurs within a circuit-controlled time window.

Within the signature analyzer is a (typically) 16-bit linear feedback shift register into which the data are entered in either true or complement logic state, according to previous data-dependent register feedback conditions. In all, with a 16-bit register, there are 65,536 possible

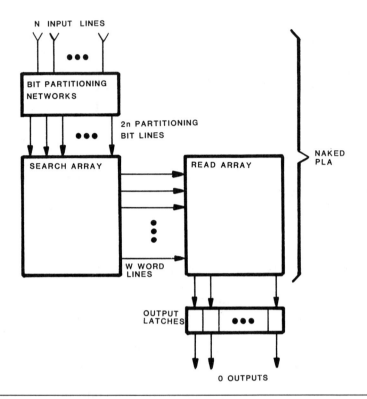

FIGURE 7-13. Logic structure not suitable for BILBO. Because their fan-in is so large, programmable logic arrays are not very amendable to BILBO-type testing.

states to which the register can be set during a measurement window. These states are then encoded and displayed on four hexadecimal indicators and become a *signature*. Each signature is then a characteristic number representing time-dependent logic activity during a specified measurement interval for a particular circuit node. Any change in the behavior of a particular node will produce a different signature, indicating a possible circuit malfunction. A single logic state change on a node is all that is required to produce a meaningful signature. Depending on the compression algorithm chosen, measurement intervals exceeding 65,536 clock cycles can still produce valid, repeatable signatures.

Serial data are shifted into the register along with start, stop, and clock signals. The remainder uniquely defines nodal states and times as long as enough patterns have been circulated through the shift register. Input stimulus vectors can either be provided by on-board software or from an external source, such as an ATE system or in-circuit emulator. See Chapter 16 for more discussion of the signature analysis technique.

REDUCED INTRUSION SCAN PATH (RISP)

While the scan (and scan-related) methodologies discussed so far offer the advantages of fully structured test generation and testing techniques, strict adherence to scan-based designs can sometimes actually increase product costs. The long test sequences required for scannable designs can increase test time, which can, in turn, limit production throughput. Strict adherence to scan-based design may also require expensive dedicated CAE tools to automatically synthesize the scan chains into a design. CAE tools are not necessarily required for scan implementation, but lack of automation tends to reduce adherence to the testability guidelines, particularly when schedule crunches occur.

Thus the many scan methodologies, just like the ad hoc guidelines, while well known and widely discussed, have not necessarily been widely adopted. Remembering the basic testability principles of partitioning, control, and visibility, it becomes apparent that a fully structured approach, while sometimes desirable, is not always necessary. This leads to the concept of *reduced intrusion scan path* (RISP) (also see Chapters 8 and 9 on boundary scan and testability busses).

With RISP, critical control and visibility nodes within the circuit under test are identified according the guidelines presented elsewhere in this text (either manually or automatically), and each critical node is augmented with a small testability cell (or T-CELL) that can be used to control and observe it. These T-cells are then connected in series, very much like a scan chain, to allow the needed communication with critical unit under test internal nodes. Only a few physical device I/O pins and very little on-chip overhead, compared to full scan techniques, are required for the testability circuitry.

A simple T-cell for testability insertion is illustrated in Figure 7-14. The NDI line is the normal data input to the node, and the NDO line is the normal data output from the node. The T-cell is inserted into and onto the node as illustrated. In normal operation, the multiplexer at the right of the illustration is set, via the TE (test enable) line, to pass NDI data straight through to the NDO line with no effect on circuit operation (except for the gate delays through the multiplexer).

For test purposes, data can be fed into the selected node by cycling the TEST CLOCK (TC), SCAN IN (SI), and SCAN ENABLE (SE) lines appropriately. Data sent into the T-cell via the SI line can then be applied to the NDO line, replacing the normal circuit's data, by enabling the TE line. For observation purposes, the SI line is enabled and NDI data are captured in the T-cell for later observation via scan-connected T-cells as shown in Figure 7-15.

As shown in the figure, the TE term can be derived from the SE and SI terms, reducing the required number of physical I/O pins to a mini-

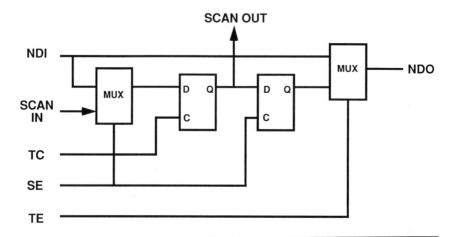

FIGURE 7-14. T-cell for testability insertion. The T-cell can be inserted in any node to partition a circuit. The T-cell can monitor the device driving the node and control the device being driven by the node.

mum of four. Thus the RISP approach can be implemented at device level with an I/O overhead no larger than that required for any other scan technique.

The potential power of the RISP approach has been demonstrated with actual circuits, as has the power of other approaches that do not insert gate delays in the circuit under test. A 20,000-gate IC, for example, which contains 900 edge-sensitive flip-flops and 35 level-sensitive flip-flops, all designed without scan, can be augmented with 38 T-cells. With less than 4 percent overhead, it is possible to automatically generate

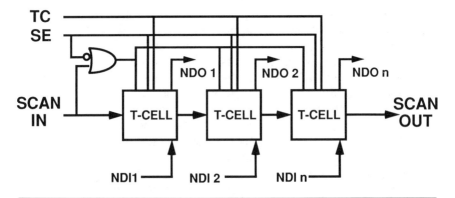

FIGURE 7-15. T-cell connections at the chip level. T-cells are connected in series to form a chain. Randomly located T-cells can then all be scanned using four package or board pins.

tests which detect 97 percent of the faults and to do so with only 16 hours of human labor and 45 hours of CPU time.

RISP, then, along with the other combination ad hoc and scan techniques discussed throughout this text, provides a viable alternative to fully structured scan designs and should be properly evaluated by each IC and board designer.

USING DEVICE SCAN PATHS FOR BOARD LEVEL TESTING

Designing and utilizing devices that include scan design of any type can make testing boards, and indeed complete systems, far easier. Scan chains can be used to break long chains of devices that are not scannable, thus partitioning a board design. They can also be used to gather data without actually being in the data path of the circuit under test and can be multiplexed to reduce the number of board level I/O pins required for good testability. Smart testers can use scan path information to make their guided probe troubleshooting algorithms work better. Combining internal scan paths (for functional IC testing) and boundary scan paths (for board level manufacturing defects testing, described in Chapter 8) can also result in lower overall business costs.

In Figure 7-16, scan paths aid in writing test programs with better fault isolation. If the scan paths were not present, bad data appearing at E or F could be caused by faults anywhere in the overall circuit. Guided probing would be a lengthy process for many faults, and a fault dictio-

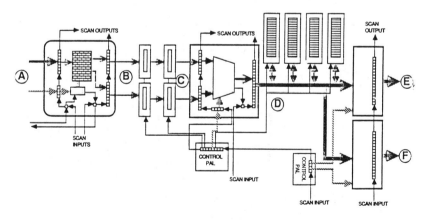

FIGURE 7-16. Breaking long device chains with scan paths. Scan devices (or scannable devices) can be placed between logic blocks to partition them and to provide control and visibility to facilitate test generation and troubleshooting.

nary would be of little value in improving this. With the scan paths, data can be tested from A to C, from B to D, and from D to E and F. This would insure that test failures would point to a specific section of the board and the amount of probing to reach a failing point would be reduced.

And even though the test would divide the board into sections, the full functionality of the board and all device interactions would still be tested due to overlapping of the sections tested. The inclusion of control PALs, or other testability circuits, in the scan path as illustrated can also aid tester control over the board. With the right testability circuits, the tester can directly inject desired control states and thus avoid lengthy control state sequencing required by normal operation of the board circuitry.

Where scan capability might degrade data throughput in a VLSI device, or on a VLSI-based board, the scan registers can be placed outside the data path to provide test visibility. This application is illustrated in Figure 7-17 using external scannable registers. Multiplexers or combination register/multiplexer circuits (for both on-line monitoring and real-time visibility and control) could also be used in the example illustrated.

In the lower left of the drawing scannable registers, along with tester interaction with the bus controller IC, allow the tester to drive and

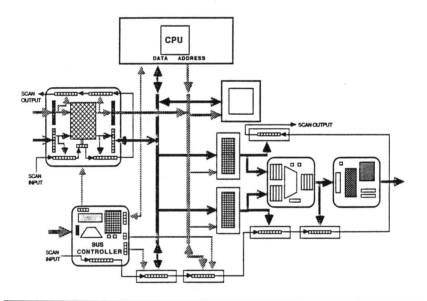

FIGURE 7-17.   Scan registers outside the data path. When placing scan registers in the data path would result in unacceptable performance penalties, they can be placed on, rather than in, the path to provide control and monitoring.

sense the data and address busses as well as to act as the bus master during at least portions of the board test. This capability can be very useful if the CPU and some of the bus devices lack scan capability. In the cluster of devices in the lower right, external scan registers are used to provide tester visibility and better fault isolation if the functional devices lack built-in scan capabilities.

It is also possible to use separate board level scan mode controls and scan data inputs and outputs to enhance tester control of a printed circuit board under test. Using separate inputs and outputs, the tester can scan different parts of the board's circuitry and provide unique input data streams for each device type. This approach is illustrated in Figure 7-18.

It is further possible to multiplex scan path outputs to allow for simultaneous scanning of devices that are located at the same places in the data path. Multiplexing of scan paths allows a more efficient functional test to be developed and executed. Proper partitioning and scan path multiplexer routing are important in a design like that in Figure 7-18. It would be a mistake, for example, to put all four of the input stage devices together on the same multiplexer instead of pairing them and routing them to both multiplexers (as shown). If all four of the input stage devices were placed on the same multiplexer, four sequential scan

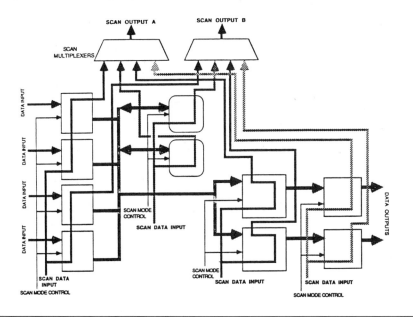

FIGURE 7-18.   Board level scan path multiplexing. Scan paths can be multiplexed to reduce the number of physical input/output pins required. This also shortens the number of test patterns applied to each scan chain.

sequences would be necessary, whereas the configuration illustrated allows all four devices to be scanned after one initialization sequence.

Board level scan paths can also make fault diagnosis more efficient, especially when using testers that combine the capabilities of a fault dictionary with a guided probe for fault isolation. Figure 7-19 illustrates how the use of scan paths, along with a fault dictionary, can greatly speed diagnostic probing by using the information in the fault dictionary to begin the guided probing sequence right in the vicinity of the actual fault on the board. While it is not always possible to include as much testability in each device as is desirable, it is always possible to consider adding as much testability to each device as is desirable (and profitable).

CROSSCHECK TECHNOLOGY EMBEDDED TESTABILITY

Another technique that can be applied to application-specific ICs is the method developed by CrossCheck Technology and licensed to LSI Logic (and others). The CrossCheck approach is unique in that it does

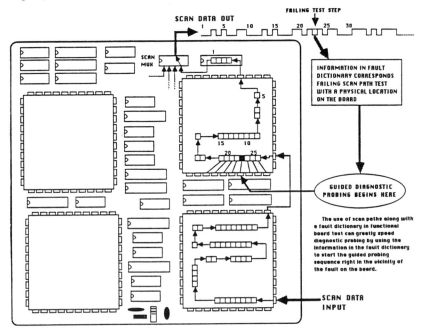

FIGURE 7-19.   Board fault diagnosis with scan paths. Some automatic test equipment can make use of scan data and a fault dictionary to start the guided probing sequence in the middle of the board (instead of at the failing card edge output pin).

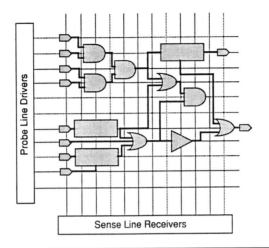

FIGURE 7-20.   CrossCheck Technology approach. The method introduced by CrossCheck allows each node in the base wafer from which a gate array is made to be observed.

not require modification of the functional logic of the IC. Instead, test circuitry is integrated into the base wafer from which an ASIC is made. The circuitry, which is conceptually illustrated for a gate array in Figure 7-20, effectively builds a bed-of-nails fixture into the base wafer of the IC.

The test architecture interlaces a matrix of sense transistors in the gate-array core. A horizontal array of probe lines controls sense transistors, which are then observable via a vertical array of sense lines. The test point itself is actually a small transistor in the gate-array base, attached to a gate's output and passing current only during testing for observation of that gate's logic state. When turned on by a horizontal probe line, each transistor in the row opens up a separate vertical sense line to the gate's state. Self-test circuitry in the periphery of the IC verifies correct gate functionally.

This approach uses up approximately 10 percent of the effective usable space of the gate array but does not require the design engineer to make any changes to his functional circuitry. There is virtually no performance degradation and the 4-pin interface, which can be accessed using lines of IEEE P1149.2 or IEEE P1149.3 proposed standard testability bus subsets, is equivalent in overhead to the boundary scan test access port package pin overhead and can be slaved to an 1149.1 TAP.

The increased visibility created by this embedded testability approach facilitates automatic test pattern generation and the overhead increases linearly (rather than exponentially, as may be the case with some scan design implementations) with the gate count of the functional circuitry under test.

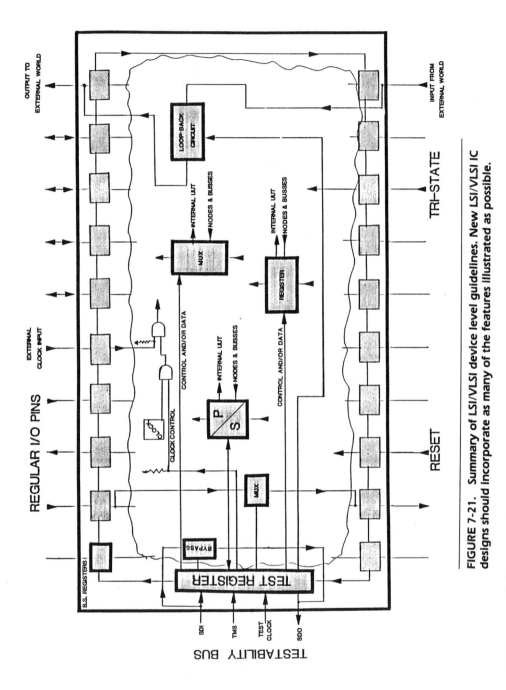

FIGURE 7-21. Summary of LSI/VLSI device level guidelines. New LSI/VLSI IC designs should incorporate as many of the features illustrated as possible.

Figure 7-21 is a summary of the recommended LSI/VLSI device level on-chip testability techniques discussed in this chapter. In addition to boundary scan, preferably compatible with the P1149 testability bus (see Chapter 10), the device illustrated utilizes internal scan registers, multiplexers, loopback circuitry, parallel-to-serial (and vice versa) converters, and external clock control. Not to be overlooked are the needs for a dedicated initialization pin (RESET) and a tristate control pin. Every attempt should be made in every case to implement as many of the testability features illustrated as possible.

If physical pin counts preclude the implementation of all of the features, on-chip circuitry can be used to decode instruction and data information from as few as four device I/O pins. Test sequence lengths will be increased, but testability will be considerably enhanced.

8

Boundary Scan

The emergence of ever more complex semiconductor devices in surface mount technology (SMT) packages and fine-pitch technology (FPT) boards and devices is putting considerable pressure on that old standby—the bed-of-nails fixture typically used with in-circuit printed circuit board assembly testers. Pin and pad spacings on boards are becoming so tight that it is literally impossible to economically and reliably build traditional fixtures for nodal access testing.

The alternative to nodal access testing—functional board testing—is also sometimes viewed as a not-so-desirable alternative because it can be expensive and time consuming to write functional test programs for complex printed circuit board designs, especially if the devices utilized are not inherently testable and if testability guidelines have not been followed during the board design phase.

Reducing the cost of functional testing to acceptable levels requires adherence to the guidelines presented in other chapters of this book. The cost of functional testing can be brought down considerably if units under test are made properly testable.

One solution available, however, in the event that the majority of board designers fail to follow functional testability guidelines rigorously, is to try to make sure that all merchant semiconductor manufacturers include circuitry within all their devices that will allow boards to be tested using the equivalent of the in-circuit testing technique without the bed-of-nails fixture.

The objective of getting circuitry that allows boundary scan of ICs at the board level is to allow for quick and simple detection of the basic spectrum of manufacturing defects (short circuits, open circuits, missing and wrong components, wrongly inserted components, etc.) without having to write functional test programs to do so. Boundary scan provides a possible alternative to the bed-of-nails fixture for manufacturing defects testing. (One might ponder the following question, however: "Why manufacture 'defects' instead of good products?")

BOARD TEST PROBLEMS AS A BASIS FOR BOUNDARY SCAN

The basic premise behind the development of boundary scan techniques is the premise that the test problem for any printed circuit board constructed from a collection of devices can be segmented into three main goals:

a. To confirm that each device performs its required function
b. To confirm that the devices are interconnected correctly
c. To confirm that the devices in the circuit interact correctly and that the complete circuit performs its intended function

At the board level, the first two goals have typically been achieved using in-circuit test techniques; for the third goal a functional test is required. With surface mount technology, however, it may not be possible to fixture for in-circuit testing. How, then, might one achieve these goals if test access is limited to the normal circuit connections plus a relatively small number of special-purpose test connections?

Considering goal (a), it is clear that the vendor of an integrated circuit used in the board level design must have an established test methodology for that component. The components are normally tested by the component manufacturer using an ATE system which may or may not take advantage of any self-test procedures embedded in the device design. Information on the test methodology used by the component manufacturer is, however, typically not available to the device purchaser.

Even where self-test modes of operation are known to exist, they may not be thoroughly documented and therefore are not reliably available for use by the component user (e.g., the board designer or board test designer). Alternative sources of test data for the board test engineer may be the device test libraries supplied with in-circuit test systems, or the test programs developed by component users for incoming inspection of delivered devices.

Wherever the test data for a component originate, the next step is to use them once the component has been assembled onto the printed circuit board. If access is limited to the normal connections of the assembled circuit, this task may be far from simple. This is particularly true if the surrounding components are complex or if the board designer has tied some of the component's connections to fixed logic levels or left pins unconnected. It will not normally be possible to test the component in the same way that it was tested in isolation unless in-circuit test is achievable. And, depending upon board level design, it may not be possible to test the component at the board level in the same way that it was tested as a stand-alone device, even given the ability to perform in-circuit testing.

To insure that manufacturing defects can be tested without requiring a bed-of-nails fixture, several companies in Europe (where the boundary scan proposal originated) and the United States decided that a framework was needed which could be used to convey test data to or from the boundaries of individual components so that their basic operation and their interconnections could be tested as if they were freestanding devices. That framework is called *boundary scan*.

BOUNDARY SCAN DESCRIPTION

The boundary scan technique involves the inclusion of latches contained in a boundary scan cell and multiplexers, one of which is in series with each component input and output physical pin. Signals at component boundaries can be controlled and observed using scan testing principles. Once these cells have been embedded in the ICs, whether merchant or customer specific, they can be used to enhance board level testability.

Figure 8-1 illustrates the basic boundary scan concept at the board level. Cells are added inside each IC between the functional logic and the physical I/O pin. These cells can be loaded and read via a scan-in line and a scan-out line (along with the appropriate test mode select and test clock inputs). By sending data in and then reading them out, board level interconnections can be tested along with several other types of manufacturing defects.

This approach works best when all of the ICs used in a board level design are equipped with the boundary scan capability. It is still workable to some degree, however, if only some of the ICs have boundary scan. An example of this situation is presented later in this chapter.

The boundary scan cells for the pins of a component are interconnected to form a shift register chain around the border of the design, and

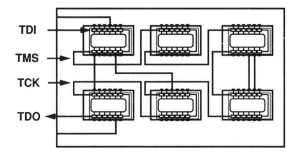

FIGURE 8-1. A boundary scannable board. Interconnection defects as well as several types of manufacturing defects (e.g., missing component, shorted pins, etc.) can be tested without a bed-of-nails fixture on a boundary scan board.

this path is provided with serial input and output connections and appropriate clock and control signals. Within a product assembled from several integrated circuits, the boundary scan registers for the individual components are connected in series to form a single path through the complete design. Alternatively, a board design could contain several independent boundary scan paths.

The diagram in Figure 8-2 shows a simple boundary scan cell. When this cell is placed between the physical input pin and the functional logic of the IC, the SIGNAL IN line is connected to the device physical pin and the SIGNAL OUT line is connected to the functional logic input. If the boundary scan cell is placed between a physical

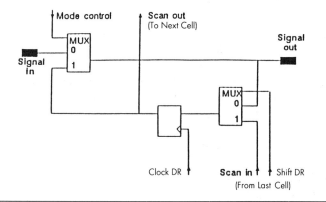

FIGURE 8-2. A simple boundary scan cell. The boundary scan cell is inserted between a device input pad and the first functional logic input or between a device's last functional logic output and an output pad on the chip.

output pin and the functional logic of the IC, the reverse is true. In any case, the structure of the boundary scan register is the same whether it is used for an input or an output.

Each boundary scan cell within an IC is connected in series to the next cell—one cell per physical digital I/O pin on the device—to form a serial scan chain that allows data to be input to the device, captured from the device, sampled from the line driving the device, or placed on the line being driven by the device.

If all the components used to construct a circuit have the boundary scan facility, then the resulting serial path through the complete design can be used in two ways:

1. It allows the interconnections between the various components to be tested. Test data can be shifted into all of the boundary scan cells associated with component output pins and loaded in parallel through the component interconnections into those cells associated with input pins connected to those output pins. This is called the *external test*.

2. It allows the components on the board to be tested. The boundary scan register can be used as a means of isolating an IC's inputs from stimuli received from surrounding components while an internal self-test is performed. Alternatively, the register can permit a limited slow-speed static test (by serially applying what used to be parallel stimulus vectors and serially evaluating what were originally parallel response vectors) of the IC's internal logic, since it allows delivery of test data to the component and examination of the test results. This is called the *internal test*.

Note also that by parallel loading the cells at both the inputs and outputs of a component and shifting out the results, the boundary scan register provides a means of "sampling" the data flowing through a component without interfering with its behavior. This sample boundary scan test can be valuable for design debugging and fault diagnosis.

The internal and external boundary scan tests allow the first two goals discussed earlier to be achieved through the use of the boundary scan register. In effect, tests applied using the register can detect many of the faults which in-circuit testers currently address, but without the need for extensive bed-of-nails access and expensive test equipment. The third goal—to functionally test the operation of the complete circuit—remains and must be achieved in the normal manner, although the ability to use sample test to examine the states of connections not normally accessible to the test system can be used to advantage.

TEST ACCESS PORT DESCRIPTION

The same four testability connection lines that are used to implement the basic boundary scan capability can also be used, with additional on-chip circuitry, to perform several additional and more sophisticated functions, as well as allowing for more sophisticated boundary scan implementations. Figure 8-3 is a block diagram of a circuit architecture known as a test access port (TAP).

The TAP contains a controller block that is actually a state machine used to configure the other on-chip testability circuitry. Depending on the state of the state machine, which is controlled by data input via the TEST MODE SELECT (TMS) line (under control of the TEST CLOCK (TCK) line), instructions can be loaded into the instruction register to configure the on-chip testability circuitry for a variety of purposes.

As shown in the drawing, a TAP can be used to control not only a more sophisticated boundary scan register but also a device identification register, one or more user test data registers (which may be imple-

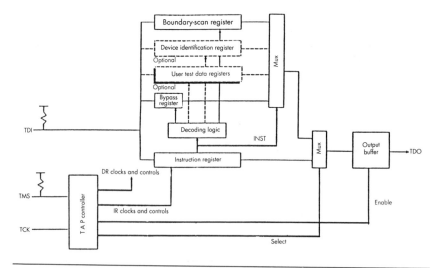

FIGURE 8-3.   Test access port. Boundary scan paths, bypass paths, user test data registers, and even device identification registers can be controlled using the test access port and a four-wire testability bus.

mented with any of the LSI/VLSI component level techniques mentioned in Chapter 7), and a bypass register.

The TAP controller generates clock and control signals for the instruction register, the bypass register, any user test data registers, and other parts of the testability circuitry. The instruction register basically controls decoding logic that in turn controls the individual data register operations and the states of the various multiplexers. The instruction sent in over the TMS line is used to select the test to be performed and/or the test data register to be accessed.

The boundary scan register allows testing of board interconnections—detecting typical production defects such as opens, shorts, and so forth. It also allows access to the inputs and outputs of devices during testing of their functional logic or to sample signals flowing through the device input and output pins.

The bypass register provides a single-bit serial connection through the circuit when none of the other test data registers is selected. This register can, for example, be used to allow test data to flow through a particular device to other devices on a board in order to speed up the testing operation by reducing the number of test vectors required to flow through each device in the scan chain when they are in the scan mode.

The device identification register is a test data register which allows the manufacturer, part number, and revision status of a component to be determined. Additional test data registers can be provided to allow access to design-specific test support facilities in the integrated circuit, such as self-test facilities.

Depending on the level of implementation and the sophistication of the TAP, the same four physical device pins that support boundary scan can also support on-chip built-in test circuitry for even the largest VLSI devices.

BOUNDARY SCAN TAP INTERCONNECTION AND OPERATION

Devices that include either boundary scan or other types of on-chip testability, as long as they contain compatible test access ports, can be interconnected in a variety of ways. Figure 8-4 illustrates a ring configuration. In the ring configuration, the TCK and TMS signals are "broadcast" simultaneously by some type of testing resource (e.g., an ATE system or an on-board BIT processor) to all of the chips under test in the ring. The TDI and TDO lines are configured in the typical scan chain serial configuration.

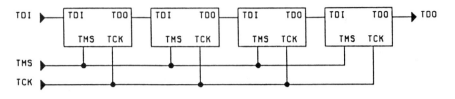

FIGURE 8-4. TAP interconnect example 1. The normal method for interconnecting devices is to connect them in series and control the resulting single scan chain with some type of bus master (usually an automatic test system).

It is also possible to configure scannable devices in a hybrid *star ring* configuration as shown in Figure 8-5. This configuration reduces the length of the test patterns required to test the components on the unit under test and reduces the size of the ambiguity group of possibly defective components in the event of a device failure. It does, however, require multiple TMS lines for its implementation.

The ultimate extension of the approach just described is to be able to have multiple test data in (TDI), test data out (TDO), and TMS lines, all controlled by a single clock, that connect to each chip. Individual star only scan chains provide the ultimate in partitioning, individual control, and visibility to each scannable chip in the design.

Boundary Scan TAP Signal Definitions

The four signals used to operate the TAP state machine have specific definitions and constraints on their operating characteristics. The following is a brief description.

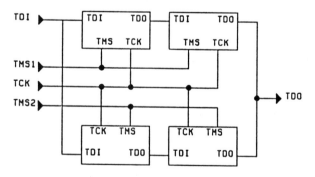

FIGURE 8-5. Hybrid star/ring configuration. Multiple scan chains can also be configured as shown. The TDO output of one scan chain is tristated via information transmitted on the TMS0 or TMS1 line while the other is active.

TCK is a free-running clock signal with a 50 percent duty cycle. The changes on TAP input signals (TMS and TDI) are clocked into the TAP controller, instruction register, or selected test data register on the rising edge of TCK. Changes at the TAP output signal (TDO) occur on the falling edge of TCK.

TDI is clocked into the selected register (instruction or data) on a rising edge of TCK. The TDI input should have a built-in pull-up resistor so that it is normally in the logic 1 state when nothing is connected to it.

TDO is the line upon which the contents of the selected register(s) (instruction or data) are shifted out on the falling edge of TCK. TDO drivers are typically set to the high-impedance state except when scanning of data is in progress.

TMS is the serial control input, and data on it are clocked into the TAP controller on the rising edge of TCK. The TMS input should also have a built-in pull-up resistor so that it is normally at the logic 1 state when it is not being actively driven.

In order to insure interoperability between boundary scannable (or other TAP equipped) devices from different manufacturers, test data received at the TDI input should appear without inversion at the TDO output following an appropriate number of rising and falling edges of TCK, regardless of the instruction or test data register selected.

Boundary Scan TAP Controller States

The TAP controller is basically a synchronous state machine. Sequencing through the various operations of the boundary scan TAP controller circuitry occurs under control of the TMS signal.

A typical main state diagram for the TAP controller is shown in Figure 8-6. The signal values shown adjacent to the state transition arcs represent the value of TMS at the time of a rising edge on TCK. A typical "subset" scan state diagram, illustrated in Figure 8-7, describes the scanning operations of the TAP instruction register (or for a previously selected test data register) once a scan operation has been ordered by data input over the TMS line. All state transitions occur based on the value of TMS at the time of a rising edge of TCK. All operations of the test logic occur on the rising edge of TCK following entry into a controller state. Logic level changes at TDO occur on the falling edge of TCK following entry into a controller state.

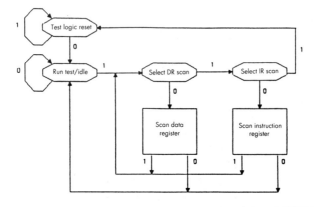

FIGURE 8-6. TAP controller main state diagram. The test access port changes state based on the sequence of 1's and 0's sent in on the TMS line (using the clock). After five logic 1's, the TAP controller returns to the test logic reset state.

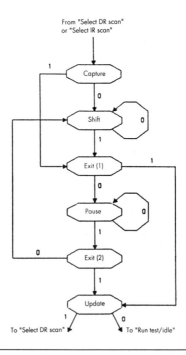

FIGURE 8-7. TAP controller scan state diagram. From the main state diagram, the TAP enters either the data register or instruction register scan block, each of which operates in the manner shown.

TYPES OF TESTS USING BOUNDARY SCAN

As mentioned earlier, three basic types of tests can be accomplished when ICs equipped with boundary scan are designed into printed circuit board assemblies: external test, sample test, and internal test. Other variations can also be accomplished for boards that are not completely scannable for accessing on-chip built-in test features and for testing busses.

External Tests Using Boundary Scan

The generalized diagram for external (i.e., interconnect) testing using boundary scan is shown in Figure 8-8 (with heavy lines showing the relevant data flow). External test was the original purpose of boundary scan, and it is well suited for it. In the example shown, data are scanned into the output pin registers of chip 1 (from the TDI line through the latches and the multiplexer) and output on its physical device pins. That data are then captured at the inputs to chip 2 and routed along the scan chain until output at device 2's TDO line.

In this way the interconnect between the devices, plus basic chip input/output functionality, orientation, and the absence of solder shorts, can be verified. If the patterns sent into the TDI line of the first

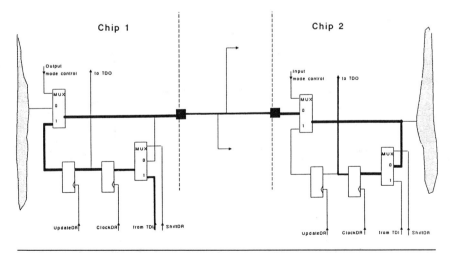

FIGURE 8-8.   External test using boundary scan. In the external test mode, a signal is sent out of chip 1 and sampled at the input to chip 2.

scannable chip are alternating logic 1's and logic 0's, it is usually possible to determine where a fault exists in the scan chain itself, since the presence of a fault would be indicated by a stream of logic 1's only or logic 0's only on the TDO line output from the last device in the scan chain. Thus using only the four boundary scan lines, given a (mostly) boundary scannable board design, the traditional in-circuit test for boundary scannable board design, the traditional in-circuit test for manufacturing defects can be performed without the need for a bed-of-nails fixture.

Sample Tests Using Boundary Scan

The boundary scan cells can also be used to sample incoming and outgoing functional (or boundary scan) data on any scannable device's input or output pins and transmit that data to a test resource on the TDO line. The signal flow for a sample operation within a boundary scannable device is shown in Figure 8-9.

At the input to the device shown, functional (or scan) data from a previous device can be latched into the boundary scan registers and subsequently output via the TDO line. If, on the other hand, it was desired to latch on-chip functional output vectors into the boundary scan cells, that can be accomplished as shown to the right of the draw-

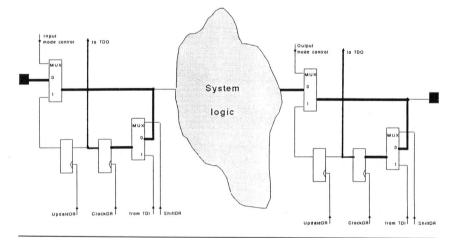

FIGURE 8-9.   Sample operation using boundary scan. The sample operation can also be used to capture the logic states resulting from normal device operation. The captured information can then be sent out the TDO line.

ing. The output sample operation can be accomplished without any effect on normal circuit operation, so a "snapshot" type of on-line monitoring (see the chapters on built-in test and testability busses for discussions of real-time on-line monitoring) can be accomplished with the boundary scan implementation.

Internal Tests Using Boundary Scan

Internal testing of the logic of a boundary scan equipped IC can also be accomplished by serializing the normally parallel-applied test vectors for a device's functional logic and sending them into the device via the TDI input line. Resulting device functional logic responses can then be captured in the device's output pin boundary scan registers and monitored via the TDO output line by reconverting the serial states captured in the boundary scan registers to parallel test vectors or by evaluating them on a serial variable-length byte-by-byte (device I/O pin size equals a byte in this case) basis. See Figure 8-10 for the signal flow for this mode.

The internal test mode is obviously most useful when a chip contains on-chip built-in test circuitry. Then the boundary scan registers need only be used to tell the on-chip testability circuitry to execute the

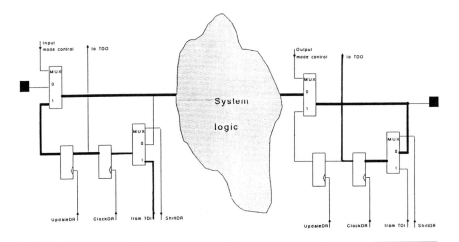

FIGURE 8-10. Internal test using boundary scan. In the internal test mode, test vectors are sent into the functional logic via the TDI line, and the resulting outputs captured and sent out on the TDO line.

built-in test routine and report the results. Trying to perform high-speed equivalents of parallel input/output test vector application/evaluation for full functional testing of the device's internal logic for all but the simplest scannable devices is obviously a tedious and time-consuming task and certainly cannot be performed at full device data rates. Where built-in test is implemented, however, the boundary scan cells do provide an additional benefit.

That benefit is illustrated in Figure 8-11, where the boundary scan mode controls are set to place the chip in the internal test mode. This mode effectively blocks incoming data from the logic driving a specific IC so that that IC can perform its own autonomous self-test. This mode of operation can only be used during either a power-on self-test or operator (or command) initiated built-in test, since it interrupts the normal function of the circuit under test and thus stops system operation (except in the case of fully redundant systems, where other duplicate circuits would continue to perform system functions while the circuit of interest is executing its self-test).

Care must be taken to include multiple TMS lines in redundant circuitry so that half of the redundant circuitry can be left in the operating state while the other half is being tested, and vice versa. Otherwise system operation can come to a complete halt. A hybrid ring star configuration should be implemented in these cases (or a simple star configuration with separate control over both the TMS and TDI lines).

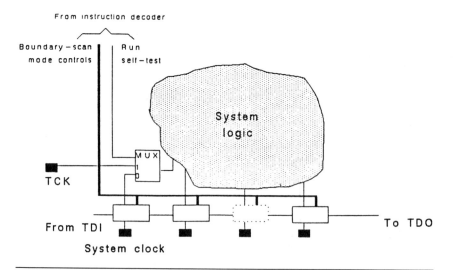

FIGURE 8-11.   Self-test using boundary scan. One mode allows the normal functional circuit's output signals to be "blocked" by the boundary scan cell so that logic states resulting from a self-test program do not propagate to other devices.

Testing Nonscannable Logic Circuits

Simple external nonscannable logic can also be tested using the boundary scan cells in scannable ICs surrounding the nonscannable logic. This operation is illustrated in Figure 8-12 where a simple two-NAND-gate latch is tested by first stimulating it (via the TDI line with the TAP in the external test mode) and then latching the results of that stimulation into the response receiving logic for later outputting on the TDO line.

This method is quite usable for simple external nonscannable circuits, but is not very practical if the external circuit is something like a VLSI processor, which takes hundreds of thousands of high-speed patterns for even a cursory test.

Testing Mixed Analog and Digital Circuits

Mixed analog and digital (i.e., hybrid) circuitry can also be tested to some extent with the boundary scan approach. Figure 8-13 illustrates an IC with digital and analog circuitry employed on the same chip. As is evident from the diagram, the analog sections of the circuitry are excluded from the boundary scan path. They are, in effect, partitioned out, as recommended in earlier chapters. One alternative to excluding the analog portions of a circuit from the boundary scan path is to implement DACs and ADCs on-chip so that analog functions can be stimulated via data input on the TDI line and the resultant (or separate) analog responses can be converted to 1's and 0's suitable for outputting on the TDO line.

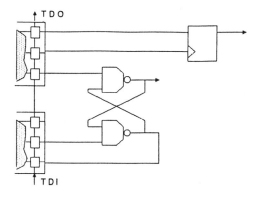

FIGURE 8-12.   Testing external logic with boundary scan. External logic can be tested by applying stimulus via the output pins of boundary scan devices and sampling the functional result of that stimulus on a boundary scan input pin.

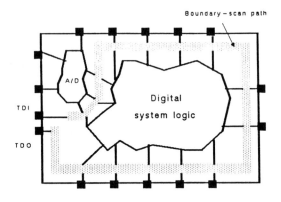

FIGURE 8-13.   Testing mixed analog and digital circuits. Boundary scan cells are placed at the partition between analog and digital circuitry in mixed signal devices.

Boundary scan techniques can also be used for testing board level bus lines as long as the proper boundary scan capabilities have been included in all of the ICs connected to the bus lines. For testing bus lines, extra boundary scan cells must be added (in addition to the cells required for the device I/O pins themselves) in order to provide separate control of tristate drivers as shown in Figure 8-14. In the example, each driver is actuated in turn, using the appropriate patterns of 1's and 0's (supplied in an interleaved fashion over the TDI line) in order to determine proper circuit operation.

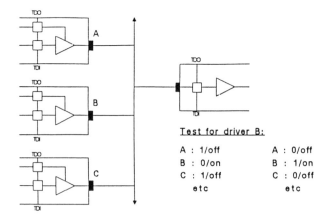

FIGURE 8-14.   Testing board level bus lines. Board level bus lines may be tested to some degree by applying the appropriate sequence of tests.

BOUNDARY SCAN CELL DESIGNS

Just as there are several variations on how boundary scannable devices can be interconnected and communicated with at the board level, there are many variations of the actual boundary scan cell that can be applied within each IC. The basic cell in Figure 8-2 provides the basic boundary scan capability, but others either provide more capability or are required when dealing with open collector and bidirectional output lines.

Figure 8-15 illustrates an input boundary scan cell design that "double buffers" data from the TDI line or to the TDO line. This boundary scan cell is actually the recommended cell design as it allows maximum flexibility in applying data to, and receiving data from, individual devices with minimum impact on circuit under test operation.

With the cell design illustrated, data input, data capture, and data output operations are separated from boundary scan shifting operations by at least one clock cycle (i.e., by at least one TAP state machine operation cycle) in order to provide less obtrusive boundary scan operations. The corresponding output cell connections (since the cell itself is basically identical) are illustrated in Figure 8-16.

Where on-chip overhead for testability circuitry is at a premium, the recommended boundary scan cells can be implemented without the output holding register. This implementation is illustrated in Figure 8-17.

Where the extra gate delays caused by the recommended boundary scan cells (either with or without output holding cells) just cannot be tolerated, input and output cells with "capture only" capability can be implemented. This type of cell is illustrated in Figure 8-18. This cell

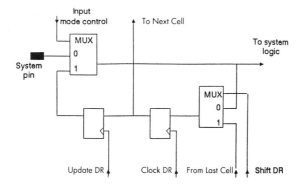

FIGURE 8-15.   Input pin cell with output holding register. Data coming from the TDI line can be shifted through to the TDO line without affecting the signal being applied to the MUX when a holding register is included in the boundary scan cell design.

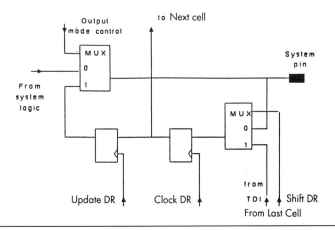

FIGURE 8-16.   Output pin cell with output holding register. Data on the system pin stay stable, even while data are being shifted on the TDI and TDO lines, until an "update DR" instruction is executed.

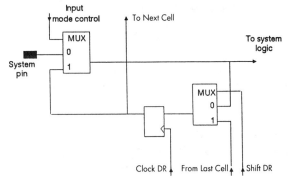

FIGURE 8-17.   Input pin cell without output holding register. Without the holding register, any shifting on the TDI and TDO lines results in shifting data being applied to the multiplexer input.

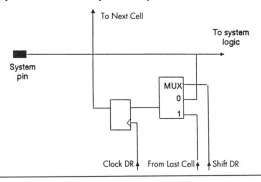

FIGURE 8-18.   Input cell allowing capture only. In circuits where the extra gate delay of the mux is not acceptable, the "capture only" boundary scan cell allows pin states to be observed but not controlled.

allows for observation of unit under test circuitry, but not for control or partitioning, and thus should only be implemented as a last resort when the recommended cells cannot be implemented.

Tristate outputs and bidirectional pins require special considerations and extra on-chip logic. The controls for the tristate and bidirectional lines each require their own boundary scan cells (in addition to the cells for the physical I/O pins themselves).

The tristate output pins, as illustrated in Figure 8-19, need extra

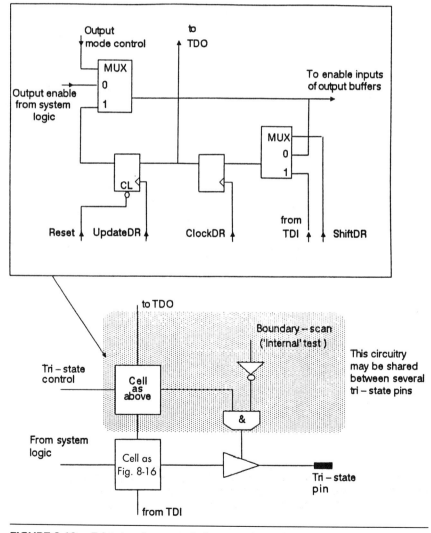

FIGURE 8-19.   Tristate pin control. An extra boundary scan cell is used to allow controlling the state of a tristate pin (or group of pins).

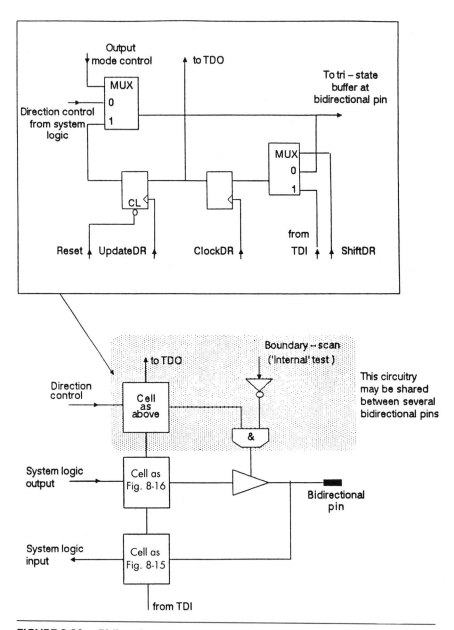

FIGURE 8-20.   Bidirectional pin control. Bidirectional pins each have two boundary scan cells—one for input and one for output—along with an extra cell to control the tristate driver (i.e., the direction control).

circuitry (which may be shared by several tristate pins) in order to isolate the device for on-chip self-test operations and to allow for bus line testing. Tristate control must be allowed both by the normal circuit tristate control input and under the control of data shifted into the boundary scan chain.

Finally, bidirectional pins require dual control of both the tristate control lines for the input/output pins and the direction control lines if boundary scan is to be properly implemented. Thus additional extra cells, as illustrated in Figure 8-20, which may be shared among many bidirectional lines, must also be included.

9

Built-In Test Approaches

Built-in test approaches offer significant opportunities for lowering the overall cost of testing both in the factory and in the field. Built-in test features in new designs reduce system integration costs, field service costs, and printed circuit board test and repair costs while providing customers with higher system availability (uptime).

Personal computers were not the first electronic systems to include built-in test capabilities, but their pervasiveness has made built-in test familiar to many people. The messages that appear at power-on (e.g., 64KB OK, 128KB OK, etc.) assure the user that the machine is still working—that is if the machine is actually testing its memory and not just printing messages!

Thinking about built-in test as a feature that customers find desirable puts a different slant on adding what is basically testability circuitry to a product design. Instead of being a burden, the built-in test features actually benefit customers and can thus increase the market share for competing products. After all, if you had a choice between an electronic system with built-in test and one without it, which would you choose? Your potential customer will answer that question just as you did.

In order to illustrate the various built-in test approaches that can be implemented, this chapter will use the system design shown in Figure 9-1. This system is composed of a central processing unit (in this case an IBM PC/AT with a VMEbus interface card that deals with the other boards in the system), various "smart" boards (boards with microprocessors on them), various "dumb" boards (boards without processors), and a variety of electro-mechanical and electro-optical interfaces to the real world. Many electronic systems are architected similarly, even though they may use different processors and busses.

Conceptual Product Configuration:

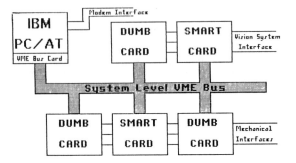

FIGURE 9-1.   Conceptual product configuration. The diagram includes a control CPU, several different circuit cards, and interfaces to The "outside world."

There are several alternatives to built-in test implementations. Each depends on the actual makeup of the system to be tested and on the goals identified for the built-in test hardware and software. Among the most common built-in test approaches are

- Centralized local built-in test
- Distributed local built-in test
- Remote diagnostic built-in test
- Field replaceable module built-in test
- Board level fault isolation built-in test
- Some combination of the above

In actual practice, it is usually the "some combination of the above" that ends up being implemented. That is because each element of the system to which BIT resources must be allocated usually has its own peculiarities and constraints. Sometimes BIT is also called built-in test equipment (BITE). Some organizations distinguish the two, so it pays to define your organization's specific terms when discussing BIT and BITE (as well as built-in self-test, or BIST).

As mentioned, built-in test implementation depends heavily on the functional configuration of the boards that make up a system. Boards (or other subassemblies) can generally be categorized in one of two ways: dumb assemblies and smart assemblies.

Dumb assemblies are typically characterized by not having any built-in processing capabilities. An example of a typical dumb printed circuit board is shown in Figure 9-2. This design is made up of gates, latches, drivers, receivers, and other glue logic, but has no on-card processor, ROM, or RAM.

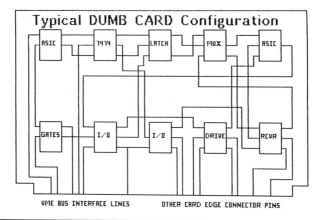

FIGURE 9-2. Dumb printed circuit board design. A dumb printed circuit board is one that does not contain a microprocessor or microcontroller and its attendant ROM and/or RAM.

Implementing BIT in a system that contains one or more of these types of board designs requires that a central BIT processor be used. And that requires adequate control over the functions of the card to be tested so that the central BIT processor can act like an ATE system by providing stimulus to the board and measuring its responses.

Control lines for things like resets, halts, and tristate conditions on the board are required if an adequate testing job is to be accomplished. Similarly, visibility lines to the various chip outputs for partitioning the board and increasing fault coverage are also needed. Successful built-in test for dumb cards requires that the built-in test features be designed into the board as its functional design progresses. The "inherent" testability features documented elsewhere in this text must be present in the design from the beginning. Atempting to "add" BIT to a circuit design that lacks inherent testability is an exercise in futility and will almost always result in a BIT fault coverage figure of well under 70 percent.

Smart boards, which are characterized by having a processor (and its associated ROM, RAM, and I/O devices), allow more flexibility and the implementation of distributed built-in test. A typical smart card design is illustrated in Figure 9-3.

A board with a processor on it can often be designed to test itself almost autonomously if the built-in test program is resident in the on-board ROM. Alternatively, the BIT program can be down loaded to the board level processor from the centralized system BIT processor. In either case, the built-in test function is distributed among the different smart boards in the system.

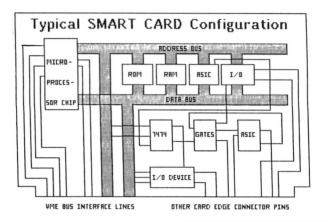

FIGURE 9-3.   Typical smart board design example. A smart printed circuit board is one that includes a microprocessor or microcontroller and the associated ROM and RAM plus other I/O devices.

Implementing high fault coverage on smart boards also requires careful attention to inherent testability features during initial board design. Control lines for chip selects and interrupts must be provided for each applicable device on the board, and status outputs and acknowledge outputs from the various devices must be visible to the BIT processor.

Systems that contain both smart and dumb board design configurations are usually tested via a combination of centralized and distributed BIT. Each method has advantages and disadvantages, but each can be used effectively if the built-in test circuitry is architected in parallel with the design of the functional circuitry.

Centralized local BIT utilizes the main processor almost like an ATE system for testing and diagnosis of boards. It requires access to all of the normal edge connector signals of the printed circuit board and to many internal signals as well. Implemented properly, centralized local BIT can allow for reasonably accurate fault isolation to a factory or field replaceable module (FRM). But it may require a large software development job and be of little or no help when it comes to diagnosis of failing components on the board itself after the board (or other FRM assembly) is removed from the system for repair.

Distributed local BIT, on the other hand, uses the main system processor only as the initiator and monitor for each individual board's BIT program. Thus it only requires access to the board's system bus lines since the on-board processor has access to the other signal lines on the card. Truly complete distributed BIT requires that each board in a system be equipped with its own processors.

With distributed BIT, the BIT software development job can be partitioned among several designers, and the presence of the on-board processor and BIT software can be of some (or even considerable) help for board level diagnosis and repair to the component level. Distributed approaches to BIT allow specialized software to be developed for each FRM in conjunction with the actual functional FRM design itself.

The BIT circuitry can be enhanced with communications capability to implement remote diagnostic BIT. This allows a technician, using a central computer at a diagnostic base location, to interrogate the system under test, force it to run its built-in test routines, and report the results. The rationale behind using remote diagnostic BIT is simple—it minimizes the number of on-site service calls that must be made and allows for the shipment of replacement assemblies (rather than the shipment of people with replacement assemblies).

Remote diagnostic BIT can be used with either centralized or distributed BIT approaches, and remote diagnostics to the exact failing FRM are quite feasible given the right amount of BIT hardware and software in the system under test. For remote diagnostic BIT to be truly effective, the diagnostic hardware and software must have the ability to control and monitor system level I/O devices as well. This allows for emulation and monitoring of system inputs and outputs.

BIT IMPLEMENTATION REQUIREMENTS

Implementing built-in test successfully requires a slightly different mind-set than that normally used in the actual system development and design. First, it requires the understanding that go/no-go type BIT is (a) not enough and (b) useful only if the system is working. Thus BIT design requires that the BIT circuitry be able to function when the system under test does not function. The usual means to achieve this requirement is to provide alternative signal paths so that faults may be isolated to the failing FRM in the absence of normal system operation.

In order to implement BIT successfully, each FRM must be designed with inherent testability (partitioning, control, and visibility). Partitioning reduces the software development effort required for the development of the BIT programs. Control allows reconfiguring of the system for fault isolation purposes. Visibility improves fault isolation resolution and allows system level diagnosis to a single FRM.

Fault isolation accuracy for BIT is a big factor. System level BIT resources require access to enough internal FRM signals to definitively determine the faultiness of a particular FRM. It also means that control and visibility must be provided to components on each FRM that may not otherwise be connected to system busses. Loopback circuitry that allows system outputs to be verified and system inputs to be emulated can provide this capability.

Finally, the BIT circuitry must be able to verify external and internal FRM signal outputs and to reconfigure each FRM (or amputate it from the system bus so that it can perform its own distributed BIT self-test without interrupting normal system functions). The normal hardware and software budget thus required for successful BIT implementation is typically between 5 and 15 percent of total system real estate.

The budgeted hardware and software space for BIT must be thought of as a product design requirement and a product marketing feature, and not as a burden or overhead just for testing purposes. The BIT budget must not be compromised in the face of demands for more features in the system. Its hardware and software space should be sacred to the design.

BIT ACCESS BUS ALTERNATIVES

There are two basic alternatives in providing a bus for built-in test access. One is to use the main system bus, or some subset of it, as the BIT access bus. The other is to implement a second bus in the system whose only job is to provide the built-in test circuitry with the needed access to each FRM.

The approach of using the system bus itself as the BIT access bus is illustrated in Figure 9-4. Testability and BIT circuitry have been added to each board in the system, and that circuitry is accessed via the system level bus (a VMEbus in the example shown).

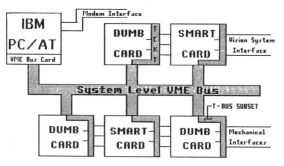

FIGURE 9-4.   BIT bus as part of system bus. When the normal system bus is also used for built-in test purposes, if the system bus goes down, the built-in test goes down with it.

Using the main system bus for the testability interface has advantages and disadvantages. The main advantage is that no extra card edge (or other dedicated test connection) pins are required in order to perform the built-in test functions. The main disadvantage is that if the system bus goes down, the BIT interface bus goes down with it. Thus the configuration in Figure 9-4 is adequate for go/no-go testing but may not be of much use for diagnostics when the system fails.

Figure 9-5 illustrates the design modifications necessary to make a dumb card testable. Visibility and control circuits have been added to the circuit board to partition it and to make internal nodes controllable and visible. The testability circuits (which may be multiplexers, shift registers, latches, or combinations of these various circuit types) interface to the main system bus. The testability circuits are treated in this case as just another peripheral device by the main system processor.

Depending on the level of sophistication one wishes to achieve when designing the testability circuits, they can be as simple as transparent multiplexers or shift registers. Or a full state machine interface, compatible with the system level bus, can be implemented to add some built-in test "smarts" to a dumb board. This approach may require the development of an ASIC for built-in test purposes, but proper design of such an ASIC could allow for its use in many different board designs.

A testable smart card example is shown in Figure 9-6. In this example, the test circuits can be accessed both by the on-board processor and by the main system built-in test processor. Also shown in this example

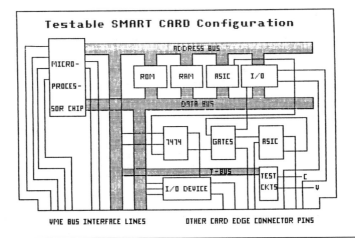

FIGURE 9-5. A testable dumb card configuration. Making a dumb card testable involves adding circuitry to it to allow for partitioning, control, and visibility from a testability bus interface of some type.

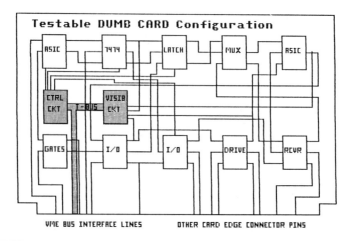

FIGURE 9-6.   A testable smart card example. Partitioning, control, and visibility in a smart card design can be accomplished by adding testability circuits that can communicate with both the on-card processor and an external processor.

is the extension of the testability bus lines down to a further level (e.g., the outside world or a dumb card controlled by a smart card) via some of the edge connector pins that are not directly accessible via the system bus interface lines.

The example illustrated in Figure 9-6 further partitions the built-in test task, allowing the processor on the smart card to do some of the work that the main system BIT processor might otherwise have to do with respect to the device being driven by the other card edge connector pins of the smart card. Many variations on the built-in test theme are possible, and there is no best way to implement built-in test. Each system must be examined in detail with respect to BIT requirements and to the proper allocation of BIT resources to each device, board, and subassembly (FRM) in the system.

To overcome the problem of losing the built-in test resources when the system ceases to function, a second port (e.g., a dedicated BIT bus) can be implemented as shown in Figure 9-7. In this example, the second port allows an alternative path for control and monitoring functions in the event of a system bus failure.

A stand-alone BIT bus (or generic testability bus) interface card has been added to the system, which can communicate via its own dedicated lines. While not shown in the diagram, the testability bus could also be extended to encompass the interfaces to the outside world, thus completely closing the built-in test loop and allowing for remote diagnostic BIT without human intervention. The inclusion of the second port improves the power of the built-in test resources beyond the go/no-

COMPLETE TESTABILITY BUS APPROACH:

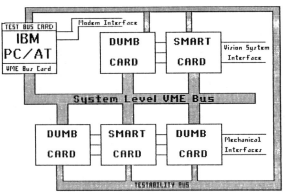

FIGURE 9-7.   Separate BIT bus implementation example. Using a separate testability bus of some type allows the main processor to use this second port to accomplish fault isolation even if the main system bus goes down.

go level and allows the system level BIT processor to definitively deal with each subassembly in the system, regardless of the ability of each subassembly to communicate with it via normal (i.e., functional) means.

CHIP LEVEL BIT IMPLEMENTATIONS

If built-in test circuitry is included in ASICs, some flexibility exists for adding on-chip BIT circuitry to the ASICs that will assist not only in testing the host chip's functional circuitry but also in testing typically nontestable or nonscannable other circuitry external to the ASIC itself. Even the most integrated designs usually include some glue logic-type functions designed using commercially available parts, and it is often these glue logic functions that are untestable via built-in test means.

Figure 9-8 shows a board design that uses two ASICs, each with its own built-in test circuitry, that has been augmented to also handle the testing of external logic.

The test circuitry inside the ASIC could include boundary scan, internal self-test or other BIST methods, as well as additional test circuitry, which could be as simple as gates, multiplexers, and shift registers, that can deal competently with the external devices (e.g., nonscannable processors, ROMs, RAMs, and other processor support and I/O chips). Burying the board level testability circuitry inside an ASIC reduces the physical board level overhead required for testability while still providing the needed BIT circuits and functions.

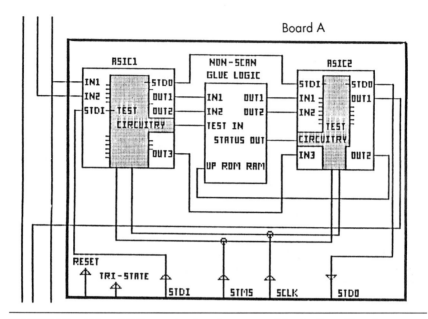

FIGURE 9-8.   Using ASIC test circuitry for external logic. Board level testability circuitry can be added inside an application specific IC to help control and observe nonscannable external logic devices.

When all of the devices in a given design are testable, preferably through some standardized scan scheme, the built-in test circuitry can be entirely contained within the individual ASICs, as shown in Figure 9-9.

BIT resources are allocated to each device on the board and are interconnected in various ways (e.g., serial only ring configurations, star configurations, or hybrid ring star configurations). Gates for testability and built-in test circuitry tend to be relatively inexpensive at the device level, while extra package pins are relatively expensive. It thus pays to make the individual devices as self-testable as possible while using as few physical package pins as possible.

At the board level, where extra devices are relatively expensive but extra I/O pins are relatively inexpensive, untestable single function devices should be replaced with dual-function devices that can provide both the performance required by the design and the testability interface needed for BIT. The required BIT circuitry can be implemented inside ASICs or by selecting commercially available testable functional parts for circuit performance design needs.

Actual complete test controller logic can also be implemented with ASIC technology, as illustrated in Figure 9-10. This figure utilizes a TAP controller (a specific testability technique requiring a state machine and

Board B

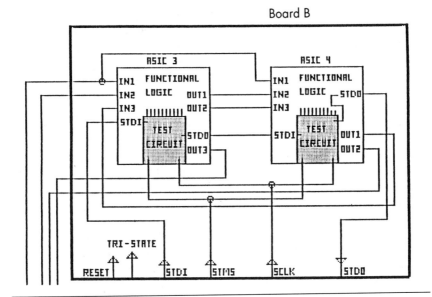

FIGURE 9-9. Built-in test with all scannable devices. When all devices are scannable, no external circuitry is required to implement testability bus communications.

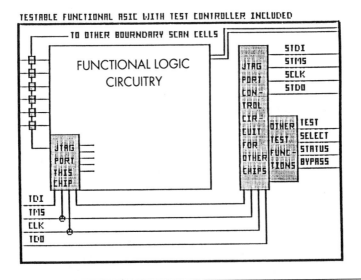

FIGURE 9-10. BIT circuitry included in ASICs. It is possible to add circuitry to an application specific IC to test the ASIC itself plus additional circuitry that is designed to test other components on the board.

formal protocol) to provide for on-chip self-test and board level interconnect and manufacturing defects testing. In addition, a second test interface circuit has been implemented on-chip to interface with other compatible circuits and noncompatible circuits. The ASIC thus designed will have more pins than an ASIC designed without extra testability features, but the overall impact on board space, and thus product size, will be absolutely minimal.

When testability circuitry is designed into a corner of an ASIC, it usually has very little impact on the function that the ASIC must perform. It also eliminates the need for adding dedicated testability and BIT circuitry to the board (or other subassembly). It is quite often possible to find enough spare gates in an ASIC design to implement at least some minimal built-in test and testability circuitry, and advantage should always be taken of those spare gates.

When utilizing serial only BIT and testability techniques, care must be taken to conform to the proper protocols. In a boundary scan environment, for example, where the bypass mode has been implemented on all devices, it is possible to address individual devices. The sequence is as follows:

1. Scan all 1's in on the test mode select line to set all devices in the mode that makes them ready to accept instruction register data via the serial test data input pin during the next scan operation.
2. Send specific patterns in on the serial test data input line to set the instruction registers of the individual devices into the desired states of operations (e.g., chips 1, 2, 4, and 5 to bypass mode and chip 3 to accept data to its user scan data registers or to the port used for external testing).
3. Send specific patterns in on the serial test data mode select input line to activate the modes entered into the individual device instruction registers via the previous serial test data input operation.
4. Send specific testing instructions (through the bypass bits of chips previously programmed for bypass mode) to the device being tested via the serial test data input line.

In a formal protocol operation, with all chips using a standard protocol, all devices in the scan chain must be first configured via a full-length scan pattern on the test mode select input data line so that the subsequent data input via the serial test data input line is interpreted properly by each device (or by extra ports included in the device that are to be utilized for testing logic external to the scan chain).

Where the number of I/O pins available for backplane or system level built-in test is minimal, a four-line serial only testability bus subset may be implemented (although six lines, which would include an initialization function and an enable function, are recommended). The example in Figure 9-11 illustrates the interconnection of the board designs shown in the previous two figures.

The example illustrated also has a feature allowing for external control of BIT bus lines in the event of the failure of any on-chip testability circuitry functions. The TEST line allows for breaking the in-system scan loop and inserting external ATE into it for controlled

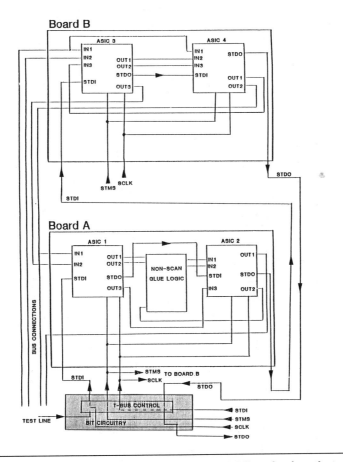

FIGURE 9-11.   BIT circuitry with external control. Adding circuitry that can switch control of the testability interface between the in-system built-in test processor and an external automatic tester provides for off-line fault isolation.

testing in the event of a failure. The TEST line can also take control of the testability bus away from the distributed BIT circuitry and command it to send and receive data from the centralized BIT processor in the system.

DUAL-PORT BIT BUS IMPLEMENTATIONS

Just as it was important at system level to provide two ports into the system for effective diagnostic BIT, it is often important to provide two ports into the BIT circuitry itself. The question "How do we test the built-in test?" often arises. The answer lies in having two ports into the built-in test circuitry itself so that it can be tested before it is activated.

Figure 9-12 illustrates the concept of a smart board utilizing a dual-port testability bus that is used for basically centralized built-in test functions (although the processor shown on the board could also utilize the combination serial and real-time addressable testability cir-

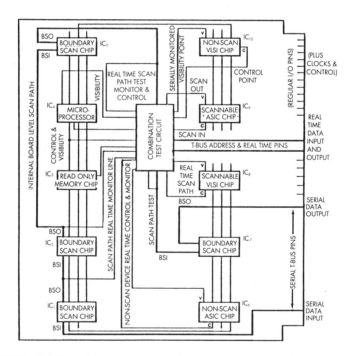

FIGURE 9-12.   Dual-port testability circuit access. When both a serial and a real-time testability interface are implemented, it is much easier to self-test the testability circuitry and to capture events in real time.

cuit interface for distributed BIT). Data can be input to the testability circuit via the serial data input lines or the real-time addressable lines. Data can then be output via the complementary lines (e.g., real-time output for serial input and serial output for real-time input) in order to verify test circuitry performance.

The wider BIT bus illustrated in the figure does use more physical I/O pins than a serial only bus, but it reduces the complexity and increases the testability of the test circuitry in the process. The combination circuits shown can deal with both scannable and nonscannable devices on boards, can provide for more definitive fault isolation within scan chains, can reduce the number of test patterns required for any given testing sequence, and can control and observe any glue logic or I/O devices that might otherwise be difficult to test from either the on-board processor or external centralized BIT processor.

The most preferred built-in test configuration utilizes a dedicated BIT processor, along with its associated ROM and RAM, to test on-board

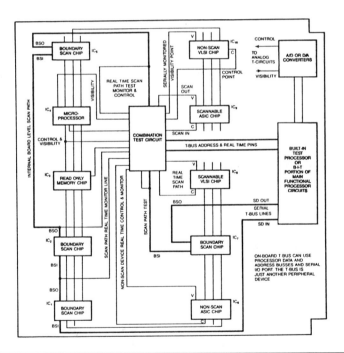

FIGURE 9-13.   Dual-port BIT bus with on-board BIT processor. Testability circuitry may be designed to look like any other peripheral device to an on-board built-in test processor that can also check analog parameters with ADCs.

(or on-chip) functions while providing BIT communications access to a centralized BIT resource in the event of functional circuit and on-board BIT circuit failures. An example is shown in Figure 9-13.

In this example, the dedicated on-board BIT processor handles testing of board functions, including analog functions, and reports results to the system level BIT processor via the serial or real-time addressable testability bus lines. If the on-board BIT processor fails, the system processor can take control of the on-board testability bus (via the TEST line) to determine exactly where a failure occurred.

The ADC and DAC functions shown in Figure 9-13 may be simple or complex, depending upon the BIT requirements. Where very accurate measurements are required, actual ADCs and DACs should be implemented. If only general conditions need to be monitored, simple analog comparators can be implemented. These comparators provide single-bit go/no-go indications to the digital testability interface. As in all BIT implementations, provision should be made for forcing errors into the analog comparators in order to verify their functionality.

BUILT-IN TEST AND HUMAN INTERACTIONS

One of the most desirable features of built-in self-test and diagnostic routines is their ability to function without (or with minimal) human interaction. Testability circuits can be used to monitor displays and to control switches that would normally require a human (see Figure 9-14A and B).

Visibility inputs are connected to the outputs of display and LED drivers to eliminate the need for human interaction (by looking at the displays) while verifying the proper performance of the functional circuitry driving the displays and LEDs. From the common testability bus, depending on the test point address selected, the built-in test circuitry can "look at" the display, the LEDs, or any of the visibility points on the functional circuit's own microprocessor.

Another human interaction problem is eliminated by connecting control point outputs (from the testability IC) to the inputs of switches. The control chip's outputs, when activated by the built-in test circuitry via the testability bus, can actually reconfigure the UUT in the same manner that a human could by closing each switch. Note, however, that if some of the switches were normally closed during board or system operation, additional gates (such as the two NAND gates used on the Motorola 68000 example) would be required so that the testability IC could "open" the switch as well as "close" it.

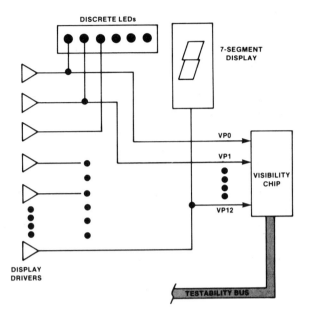

FIGURE 9-14A. Visibility to eliminate interaction. Logic driving displays can be monitored electronically to eliminate the need for human observation of what may be many sequences of the display.

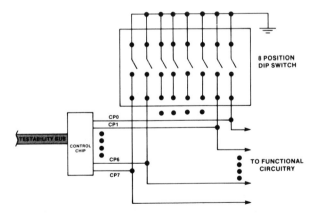

FIGURE 9-14B. Control to eliminate interaction. Logic controlled from a testability bus can be used to emulate the operation of switches, thus allowing for real-time reconfiguration of the unit under test without human intervention.

REAL-TIME ON-LINE MONITORING

Using a testability chip set as the interface between the testability bus and the internal nodes of the functional circuitry allows for real-time monitoring, on a visibility point by visibility point basis, of any signal to which a testability point has been connected. The built-in test processor might, for example, want to periodically compare actual data from the UUT internal node with "expected" data stored in a test point data ROM and light an error light (or take other action) if a mismatch occurs, indicating a failure in the functional circuitry under test.

Any critical node in the UUT can be selected for connection to a testability circuit visibility point input during the design phase. The scan-in and scan-out lines on application specific semicustom ICs, for example, are prime candidates for control and visibility point connections. Access to these lines, which were included in the device design to render it testable, is even more critical once the IC has been installed on the printed circuit board. Having access to the individual chip self-test inputs and outputs makes fault isolation much more accurate than if all of the scan lines were simply connected in series.

In summary, built-in test approaches can take a wide variety of forms. The key guidelines to success are enough internal partitioning, control, and visibility of each FRM to definitively determine goodness or badness and two ports in both the functional and testability circuitry (three ports total) so that the main system BIT processor can perform testing and fault isolation even when the main functions of the system under test are not working.

10

Testability Busses

There are really only four general approaches to providing improved testability, three of which have already been discussed (serial, ad hoc, and probe-ability implementations). The fourth is the testability bus (T-Bus) approach.

Serial approaches are most popular at the device (IC) level, since they use very few device I/O pins. When the I/O pin restriction is lifted at the board level, ad hoc, or probe-ability approaches are often used. The T-bus is designed to allow circuit designers to select and combine the optimal combination of testability techniques from those available, while simultaneously providing test personnel with a standard interface to whatever testability circuitry is implemented.

Serial approaches use scan-in, scan-out, and clock lines to shift data into and out of the testability interface of the UUT. At the device level, implementing scan design means using a highly structured approach to logic design and typically using from 3 to 20 percent (or more) of the available silicon. At the board level, the scan lines may be cascaded (i.e., connected in series) so that, if protocols match, the scan path can be used for multiple devices on the board.

The ad hoc testability approaches are individual attempts to make sure that devices, boards, and systems are initializable, controllable, and observable. Because there are a lot of alternatives, applying ad hoc guidelines sometimes requires many discrete test connections. These connections may be difficult to interface to physically at the board level and may impact UUT circuit functional performance, depending again on how they are made physically accessible. A standard testability bus

---

* The information in this chapter was accurate at the time of printing. The DoD and IEEE standards and proposed standards are subject to change, evolution, and refinements. Refer to the appropriate DoD and/or IEEE P1149.x series of specifications for the latest information.

is designed to solve this problem by providing relatively few signal lines that can be used to access many UUT internal functional nodes.

With the increasing density and complexity of printed circuit board assemblies, particularly those making use of surface mount technology (SMT), some designers have adopted the *probe-ability* approach. With this method, devices with closely spaced leads are mounted on the board, and discrete test pads are placed around the devices so that they can be probed with a bed-of-nails test fixture.

Fanning out closely spaced leads so that they can be accessed mechanically requires 10 to 30 percent of available board space, depending upon component configurations and whether or not some components are restricted to only one side of the board. The close center, sometimes dual-sided, fixtures used to probe SMT boards are expensive and not as reliable as is desirable. They may also significantly affect functional circuit performance when they make contact with the board under test due to added capacitive loading and crosstalk within the test fixture.

The T-bus, in contrast, is an approach that overcomes virtually all past problems, especially at the board and system levels, and will support future design methods, technologies, and complexities. The T-bus provides a standard testability interface that supports both structured (scan) designs and ad hoc methods in any combination.

The T-bus is a combination analog/digital, serial/parallel bus that allows for real-time data input and output to any addressable control or visibility point, while at the same time supporting all existing and proposed serial-only approaches. If edge connector pins are unavailable for the T-bus interface, the T-bus can be brought to easily probe-able pads for physical access. Using the T-bus thus allows a designer to implement the optimal amount of partitioning, control, and visibility with minimal parts selection and logic design constraints and with minimal added parts cost or board and system real estate (space) penalties.

Implementing the T-bus is done either by adding one or two test circuits to a design to partition it or by replacing an untestable circuit component (a shift register, multiplexer, addressable latch, etc.) with a testable component that has input and output pins for circuit functions and the needed testability functions. The testability functions are then interfaced to the T-bus.

At the board level, for example, functional nodes that need control on the UUT are driven by the testability circuits from the T-bus (see Figure 10-1). Similarly, critical nodes that need to be observed are connected to testability circuit visibility point inputs and can be

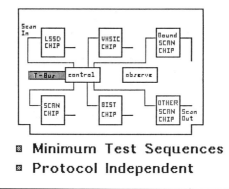

■ Minimum Test Sequences

■ Protocol Independent

FIGURE 10-1. Partitioned control approach. Control circuits added to a design help to partition it and provide control of a wide variety of different devices with virtually no impact on circuit performance.

monitored via the T-bus without interfering with normal functional circuit operation (see Figure 10-2).

Combining control and visibility testability circuits (see Figure 10-3) allows test generation with the fewest test sequences, and provides for unambiguous fault isolation for nonscannable and scannable (using any scan protocol) devices on the board.

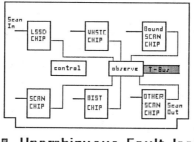

■ Unambiguous Fault Isolation

■ Protocol Independent

FIGURE 10-2. Partitioned visibility approach. Visibility circuits added to a design help to partition it and provide visibility to a wide variety of different device outputs with virtually no impact on circuit performance.

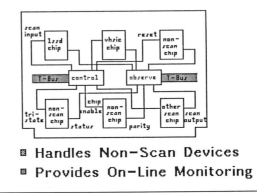

☒ Handles Non-Scan Devices

☒ Provides On-Line Monitoring

FIGURE 10-3.   A universal approach. Combining control and visibility circuits in a design provides partitioning, control, and visibility through a simple testability bus interface.

THE PROPOSED IEEE STANDARD TESTABILITY BUS

A group within the Test Technology Technical Committee of the Computer Society of the IEEE is working toward official standardization of a series of testability bus descriptions for analog and digital devices, boards, and systems. This committee is known as the P1149 Testability Standards Steering Committee (TSSC). The TSSC is made up of several groups. The P1149 group is responsible for creation and approval of an overall guide to testability bus implementation. The technical working group subcommittees (P1149.1 through P1149.x) are responsible for generating specific protocol and implementation documents that describe specific aspects of the various subsets of the overall testability bus. The overall group recognizes that a full testability bus is the ideal situation but also realizes that it may not be possible to implement the complete bus in all cases. The proposed P1149 standard testability bus guide, therefore, also allows for subsets of the full bus where input/output pin constraints cannot be overcome. If the subsets conform to the standard, however, they become part of an overall testability system (see Figure 10-4) that is usable at any level with any type of testing approach.

When selecting components for a new design, it is recommended that, where a choice exists between a component that does not include testability features and an equivalent component that does include testability features, components that include testability features be selected. When designing new full custom or semicustom ICs, it is recommended that the design include testability features. These testability features may include LSSD-like scan chains to partition the

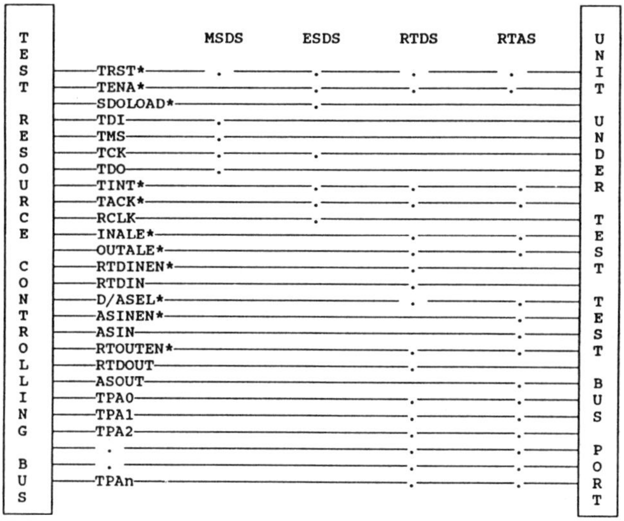

FIGURE 10-4. The originally proposed IEEE standard testability bus. The originally proposed IEEE P1149 standard testability bus included both digital and analog capabilities and both serial and real-time features for the digital portions.

functional circuitry and to make it controllable and observable. As a minimum, however, a boundary scan-type structure should be included in each new IC design to facilitate testing and testability bus implementation at the board level.

This chapter presents examples of schematics and block diagrams of the various types of testability circuitry that can be included in a UUT design to implement all or part of the testability bus. Each testability implementation technique has its own advantages, disadvantages, and capabilities and must be evaluated by the circuit designer so that the most appropriate technique or combination of techniques for each specific application are used to implement the P1149 testability bus (or the appropriate subsets with the appropriate protocols).

Figure 10-5 illustrates the hierarchy of IEEE P1149 documents as they relate to the overall testability bus. The proposed standard bus was divided into four discrete subsets:

- Minimum serial digital subset (MSDS)
- Extended serial digital subset (ESDS)
- Real-time digital subset (RTDS)
- Real-time analog subset (RTAS)

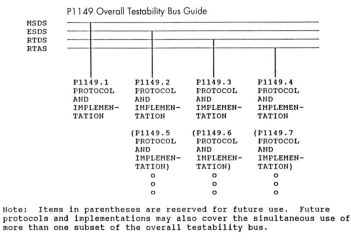

Note: Items in parentheses are reserved for future use. Future protocols and implementations may also cover the simultaneous use of more than one subset of the overall testability bus.

FIGURE 10-5. Hierarchy of P1149 T-bus documents. The overall P1149 document describes the interrelationships between the various "dot" documents, each of which specifies a particular implementation and protocol. Items in parentheses are reserved for future use. Future protocols and implementations may also cover the simultaneous use of more than one subset of the overall testability bus.

Changes in the P1149 structure since its inception have resulted in the abandonment of the P1149.4 Real-Time Analog Subset of the testability bus. IEEE-Std-1149.1 has been approved as of this printing, and work continues on P1149.2 and P1149.3.

Fully structured and semistructured (ad hoc) digital and analog testability techniques, some MSDS and some ESDS, some RTDS and some RTAS, and some full P1149 testability bus combination serial/real-time examples, are presented in this chapter. They are by no means inclusive but are presented as applications guidance to those implementing the testability bus.

As much testability as it is possible to implement should be included in each new circuit design. Testability is a design practice which utilizes ad hoc guidelines and structured techniques to ease prototype/production testing and field maintenance/repair. It balances the performance/reliability trade-offs against the total product cycle costs to achieve an effective, efficient, and economic solution.

Testability may also be defined as a design characteristic which allows the status (operable, inoperable, or degraded) of an item to be determined and the isolation of faults within the item to be performed in a timely manner.

TESTABILITY BUSSES AND LSSD

As mentioned in the first section of Chapter 7, LSSD is IBM's discipline for structured design for testability. The proposed IEEE standard testability bus can interface directly with LSSD and LSSD-like circuits as long as the proper protocol is followed on the testability bus lines.

P1149 ESDS testability bus subset line equivalents to the LSSD input/output lines are shown in Table 10-1. LSSD circuits are directly compatible with the P1149 ESDS signal lines and the P1149 RTDS real-time data signals, properly decoded by on-board testability circuitry, and can also be connected to individual or cascaded LSSD scan inputs and scan outputs when TCK and RCLK are in the correct states.

TABLE 10-1.   P1149 ESDS to LSSD Equivalent Lines

P1149 ESDS Line	LSSD Line
TDI	Scan data in
TDO	Scan data out
TCK	Clock input A
RCLK	Clock input B

TESTABILITY BUSSES AND BOUNDARY SCAN

Boundary scan signal lines and protocols are directly, functionally, and protocol compatible with the P1149 MSDS of the testability bus. The P1149 MSDS equivalent signal lines for implementing boundary scan are shown in Table 10-2.

TABLE 10-2.   P1149 MSDS to Boundary Scan Equivalent Lines

P1149 MSDS Line	Boundary Scan Line
TDI	TDI
TDO	TDO
TCK	TCK
TMS	TMS

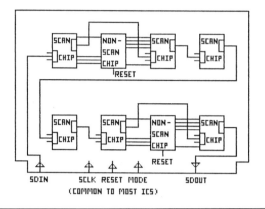

FIGURE 10-6.   Adding TRST* to the P1149 MSDS. One option allowed by the P1149.1 standard is the option of adding a fifth line, TRST*, which is used to directly (and asynchronously) initialize the test access port without a potentially long serial data stream.

It is strongly recommended that ICs selected from commercial IC manufacturers for new designs include, as a minimum, a basic boundary scan capability that is 1149.1 protocol compatible, and that new in-house designed ICs also include such a capability. Where both scannable and nonscannable ICs are included in a design, it is recommended that the expanded serial digital P1149 testability bus subset be implemented at the board level. Two examples of P1149 ESDS implementations are illustrated in Figures 10-6 and 10-7. In Figure 10-6, the testability bus TRST* line (P1149 ESDS) has been added in order to directly initialize any ICs that cannot be easily set to known states via

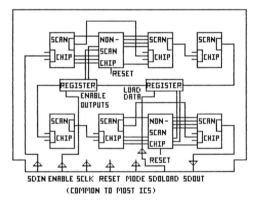

FIGURE 10-7.   Complete P1149 ESDS example. Using the complete extended serial digital subset allows for direct control of many types of testability circuitry, including scan path, scan/set, and LSSD.

the scan path. In Figure 10-7, the complete P1149 ESDS subset has been implemented using additional shift registers to control nonscannable IC inputs and to observe nonscannable IC outputs in order to partition the circuit more readily, to reduce the number and complexity of test patterns that must be generated, to improve fault isolation in the scan chain, and to allow slower test resources to evaluate data loaded in real time (via the SDOLOAD* testability bus line) at a later time. The TRST* line is connected to the nonscannable ICs requiring initialization and to the registers added for control and visibility purposes.

TESTABILITY BUSSES AND SCAN/SET

The scan/set testability implementation is illustrated in Figure 10-8. With this approach, UUT functional circuitry is modified only insofar as is required to allow the parallel outputs of the scan/set register(s) to control functional circuit inputs and the parallel inputs of the register(s) to load data from UUT visibility (observation) nodes for later shifting out on the serial data output line.

Scan/set circuitry, properly designed, can be fully supported by the P1149 ESDS subset of the testability bus. Care must be taken to insure that proper logic states and clock edges are selected when designing the circuitry. Table 10-3 lists the P1149 ESDS line equivalents for scan/set implementations.

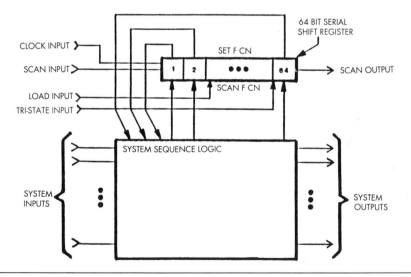

FIGURE 10-8.   Scan/set testability implementation. A 64-bit shift register with tristate outputs is used to illustrate the connections required from the testability bus.

TABLE 10-3.   P1149 ESDS to Scan/Set Line Equivalents

P1149 ESDS	Scan/Set Line
TENA*	Tristate outputs*
TRST*	Master reset*
TDI	Serial data input
TCK	Shift clock
TDO	Serial data output
SDOLOAD*	Parallel loan input

TM AND E-TM TESTABILITY BUSSES

There are several testability circuit and testability bus-like techniques being implemented in VLSI designs in addition to the standard functional and boundary scan techniques. This section discusses the TM and E-TM busses developed as part of the very high speed integrated circuit (VHSIC) program sponsored by the U.S. Department of Defense (DoD). These techniques use combinations of scan techniques and various built-in test and built-in self-test circuits that can be used at the IC and board levels. Some can be further extended to the subsystem level, and all can be interfaced to using combinations of the P1149 MSDS and/or P1149 ESDS testability bus subsets.

TM-Bus Implementation

The TM-bus is a four-line bus with an independent clock source that uses the concept of master and slave testability interface circuits, as illustrated in Figure 10-9. The TM-bus uses an elaborate and highly

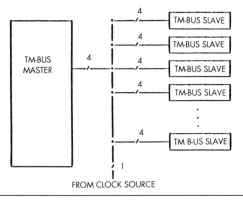

FIGURE 10-9.   TM-bus conceptual model. A TM-bus implementation has one master controller and up to 32 slaves, all driven from a common clock source.

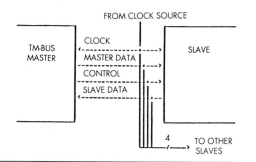

FIGURE 10-10.   TM-bus signal names. The master data signal is the serial data input signal to the slaves, the slave data signal is the serial data output, and the control signal is the mode selection input.

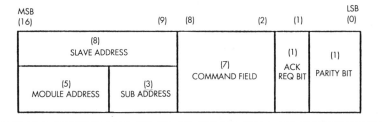

FIGURE 10-11.   TM-bus header packet. All messages along the TM-bus have the same format. The header packet tells a slave that it is being addressed and what to do with the rest of any subsequent message bits.

TABLE 10-4.   P1149 to TM-Bus Equivalents

P1149 MSDS/ESDS Line	TM-Bus Line
TCK	Clock
TDI	Master data
TMS	Control
TDO	Slave data

specified serial data protocol in order to perform multiple testability functions over the fewest dedicated testability lines. Figure 10-10 shows the lines that make up the TM-bus. Figure 10-11 shows the structure of the header packet that is first sent down the serial data input line from the master to the slaves, while Figure 10-12 shows the state diagram for the circuitry in each slave.

The TM-bus signal line functions are supported by the P1149 MSDS subset of the testability bus, but the protocols on the signal lines require further specification. Testability bus lines that cannot be implemented with the P1149 MSDS subset can be driven and monitored by using the P1149 real-time digital subset of the testability bus. The equivalent P1149 MSDS signal lines for the TM-bus are shown in Table 10-4.

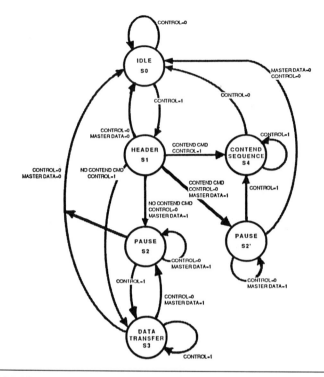

FIGURE 10-12.   TM-bus state diagram. Implementing the TM-bus requires a test circuitry controller with a state diagram.

E-TM-Bus Implementation

The E-TM-bus is another approach to a serial path for test and maintenance control and data information at the IC (or chip) level. The E-TM-bus consists of six lines and, like the TM-bus, uses the master/slave concept. It is intended to interface to the TM-bus via a master/controller,

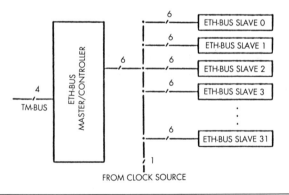

FIGURE 10-13.   E-TM-bus conceptual model. The E-TM-bus extends the range of the TM-bus by translating the four-line TM-bus to the six-line E-TM-bus.

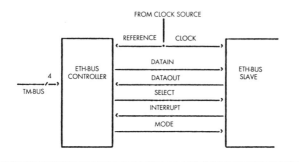

FIGURE 10-14. E-TM-bus signal types. There are six signals associated with the E-TM-bus. The SELECT and INTERRUPT lines are the additions to the normal CLOCK, MODE, DATA IN, and DATA OUT lines.

as illustrated in Figure 10-13. As shown, the E-TM-bus is configured to interface one master controller with up to 32 slaves.

The individual signal lines for the E-TM-bus are shown in Figure 10-14.

The E-TM-bus was designed to allow individual testable circuits to be connected in either a ring or a star configuration, as illustrated in Figures 10-15 and 10-16.

A conceptual view of the logic required to implement the E-TM-bus in an IC or on a board is presented in Figure 10-17. The logic for the E-TM-Bus interface is customized for each element (i.e., VLSI circuit) in order to render it as testable as possible while still providing a standardized interface and protocol.

The P1149 ESDS subset of the testability bus supports the functions of the lines of the E-TM-bus, but the protocol must be further specified. The real-time subset lines of the P1149 testability bus can be used to support E-TM-bus protocols in conjunction with the P1149 ESDS lines. The nearest equivalent P1149 ESDS lines are shown in Table 10-5.

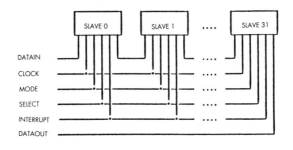

FIGURE 10-15. E-TM-bus ring bus structure. E-TM-bus–compatible circuits can be configured so that the data output from the first one feeds the data input of the next one and the data out signal is acquired from the last device in the chain.

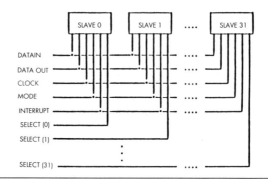

FIGURE 10-16. E-TM-bus star bus structure. E-TM-bus–compatible circuits may be configured so that they share the common bus lines with the exception of the SELECT line, which is then used to determine which device is active.

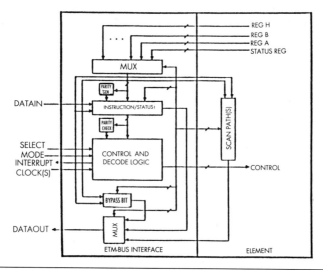

FIGURE 10-17. E-TM-bus interface concept. The on-chip test access port for the E-TM-bus includes control and decode logic, instruction/status registers, a bypass bit, user test data registers, and multiplexers.

TABLE 10-5. P1149 ESDS to E-TM Bus Closest Equivalents

P1149 ESDS Line	ETM-Bus Line
TENA*	Select
TDI	Datain
TDO	Dataout
TMS	Mode
TCK	REFCLK
TINT*	INT

TESTABILITY BUSSES AND THE TAP

As a result of considerable research, development, communication, and cooperation among many of the parties to the P1149 testability bus effort, a very strict protocol for, and implementation of, the four-line minimum serial digital subset of the P1149 testability bus, described in the 1149.1 test access *Port and Boundary Scan Architecture* document, has been defined that supports both the boundary scan architecture for manufacturing defects testing and many chip level testability and built-in test (including built-in self-test) functional test approaches. This implementation of the four-line P1149 MSDS of the testability bus utilizes a testability controller architecture called a test access port (TAP). The block diagram for this port is illustrated in Figure 10-18.

The test access port utilizes four lines that are exactly equivalent in function and logical relationships to the minimum four-line P1149 testability bus subset. Its internal circuitry decodes the serial data inputs to the TDI and TMS lines in order to configure the UUT functional circuitry for normal operation, boundary scan operation, and internal and external test operations. The specific protocols for TDI and TMS line inputs provide for simplified entry into the bypass and boundary scan modes and are fully defined, including specific implementation examples and detailed circuit designs, in the IEEE 1149.1

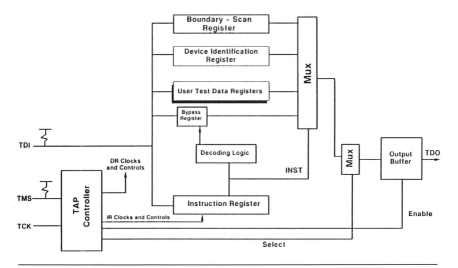

FIGURE 10-18.   IEEE-Std-1149.1 test access port architecture. The test access port includes a state machine (the TAP controller) to decode instructions arriving on the TMS line.

document entitled *Standard Test Access Port and Boundary-Scan Architecture*.

When using the T-bus to support the P1149 MSDS, it is important to remember that the P1149 real-time subset of the testability bus can be used at the board, subsystem, and system levels in combination with the P1149 MSDS and/or P1149 ESDS subsets to support multiple boundary scan chains in order to improve board partitioning and achieve more rapid and accurate fault isolation to a failing component while reducing test pattern generation and test times.

REAL-TIME TESTABILITY BUSSES AND MULTIPLEXING

The P1149 real-time digital subset of the testability bus may be implemented with multiplexers and/or decoders added to the functional circuit. Figure 10-19 shows one implementation of the minimum P1149 RTDS using simple tristateable multiplexers and decoders without address latching capability.

Figure 10-20 shows an expanded example, utilizing all of the lines of the P1149 RTDS of the testability bus. This example takes advantage of the INALE* and OUTALE* lines so that data can be simultaneously input to and output from analog and digital portions of a circuit. Implementing the INALE* and OUTALE* lines lends considerable flexibility in controlling and observing critical UUT nodes in real time.

MINIMUM RTDS TESTABILITY BUS

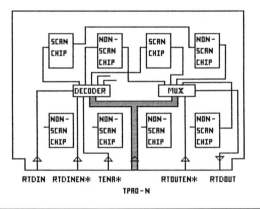

FIGURE 10-19.   Minimum P1149 RTDS multiplexer implementation. The real-time lines of the P1149.3 testability bus can be implemented using simple multiplexers and decoders. Address lines determine which nodes will send and receive testability data.

EXPANDED RTDS TESTABILITY BUS

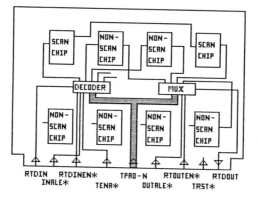

FIGURE 10-20.   Expanded P1149 RTDS multiplexer implementation. More sophisticated testability circuit designs which make use of the address latch enable and handshaking capabilities of the testability bus are also possible.

The multiplexing approach provides a means of also inherently partitioning a circuit to enhance its fault isolation characteristics and to reduce the logic simulation and test programming effort required for high-quality fault coverage.

COMBINATION SERIAL/REAL-TIME TESTABILITY BUS

One of the most effective approaches to implementing the P1149 testability bus is to take advantage of both the P1149 MSDS or P1149 ESDS and the P1149 RTDS simultaneously. This implementation method provides for inherent partitioning, control, and visibility, helps achieve the highest-quality test with the fewest test patterns (and therefore the least effort needed to generate and verify them), and enhances fault diagnostic resolution. Using both subsets also provides two ports into the functional circuitry of the UUT and the testability circuitry added to it (or substituted for typically untestable single-function circuits). Two ports allow flexibility in test pattern application, interfacing to UUT nodes (scannable and nonscannable), and easy self-testing of the testability circuitry itself.

Figures 10-21 and 10-22 illustrate adding parts to a printed circuit board assembly for testability purposes. Each of the circuits added, called control and visibility circuits, has the capability of interfacing any UUT internal node to either (or both) the serial input/output lines or the real-time digital input/output lines of the testability bus.

The control circuit can perform virtually all of the functions supported by the testability bus. A combination serial/real-time control

CONTROL CIRCUIT SERIAL/REAL TIME DATA INPUT

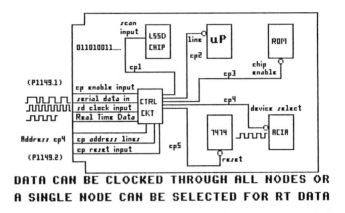

DATA CAN BE CLOCKED THROUGH ALL NODES OR
A SINGLE NODE CAN BE SELECTED FOR RT DATA

FIGURE 10-21.   Control circuit implementation. A control circuit that contains both registers and real-time address decoding circuits can be added to a design to provide the drive functions of a dual-port testability bus interface. Data can be clocked through all nodes, or a single node can be selected for real-time data.

VISIBILITY CIRCUIT SERIAL/REAL TIME DATA OUT

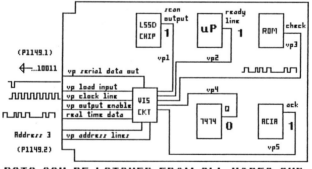

DATA CAN BE LATCHED FROM ALL NODES AND
SHIFTED OUT AND ANY NODE MONITORED IN R-T

FIGURE 10-22.   Visibility circuit implementation. A visibility circuit that contains both registers and addressable multiplexers can be added to a design to provide the monitor functions of a dual-port testability bus interface. Data can be latched from all nodes and shifted out and any node monitored in real-time.

circuit should be used in either the serial mode or the real-time mode at any one time, not in both modes simultaneously.

The visibility circuit can perform both serial and real-time data operations simultaneously and provide on-line monitoring capabilities.

It is not always necessary to add dedicated circuitry to implement the P1149 testability bus. Many times, functional glue logic circuits can be replaced with testable functional circuits that can perform the required circuit function and provide the testability bus interface capability. Testable functional circuits have a few more pins than typically untestable single function circuits, but can, in general, provide the testability bus interface, either P1149 MSDS/P1149 ESDS or P1149 RTDS, with an absolute minimum of overhead.

Figure 10-23 illustrates replacing typical multiplexer and/or decoder circuits with testable functional circuits that also include a shift register function that interfaces with the P1149 ESDS portion of the testability bus. The multiplexer and/or decoder functions contained in the circuits are used to implement the required logic function for UUT performance. The extra serial port is used to implement the P1149 MSDS and/or P1149 ESDS subset portions of the T-bus.

Figure 10-24 illustrates replacing typical shift register functions with testable functional circuits that also include decoder and/or multiplexer circuits that interface with the P1149 RTDS portion of the testability bus. The shift registers contained in the circuits are used to

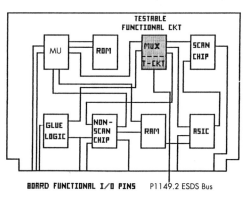

TESTABLE MULTIPLEXERS AND DECODERS, ETC., REPLACE SINGLE FUNCTION
CIRCUITS TO PROVIDE BOTH FUNCTIONALITY AND A TESTABLITY BUS INTERFACE

BOARD FUNCTIONAL I/O PINS   P1149.2 ESDS Bus

FIGURE 10-23.   Testable functional circuits for P1149 MSDS/ESDS. It is not always necessary to add dedicated testability circuits. Testable functional circuits use part of the circuit for circuit function and the other part as a testability interface.

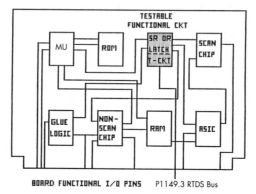

FIGURE 10-24. Testable functional circuits for P1149 RTDS. Where a shift register or latch function is needed, an extra real-time port can provide testability control and observation capabilties.

implement the required logic function for UUT performance. The extra real-time port is used to implement the P1149 RTDS portion of the T-bus.

ANALOG TESTABILITY BUS IMPLEMENTATION

The originally proposed P1149 analog real-time subset of the testability bus, while not currently being pursued as an IEEE standard, can be easily implemented with analog switches (decoders) and multiplexers.

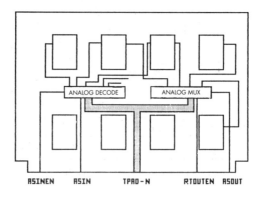

FIGURE 10-25. P1149 real-time analog minimum subset implementation. Analog multiplexers and decoders can be added to a board and addressed via the P1149 address lines. Analog signal information is input and output in real time under the original proposal.

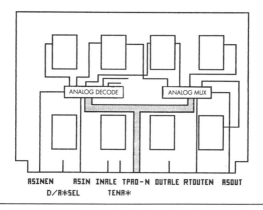

FIGURE 10-26. P1149 RTAS expanded subset implementation. Use can be made of the ADDRESS LATCH ENABLE and D/A* SELECT lines to design relatively sophisticated combination digital and analog testability bus interfaces.

Figure 10-25 illustrates one implementation of the minimum configuration of the P1149 RTAS, while Figure 10-26 shows how the complete P1149 real-time analog subset can be implemented using extra circuitry to take advantage of the D/ASEL* and INALE*/OUTALE* lines. Using the full capability of the P1149 RTAS, especially when other subsets are also implemented, provides considerable flexibility in getting digital and analog testability information into and out of the UUT with the fewest test sequences and the least overhead circuitry.

TESTABILITY BUS CONFIGURATION OPTIONS

The originally configured P1149 testability bus can be implemented, wholly or in parts, at any level of circuit integration. It is also amenable to built-in test and built-in self-test applications at board, backplane, subsystem, and system levels. Figures 10-27 through 10-30 show some of the possibilities for implementing the testability bus for various system configurations.

In Figure 10-27, the P1149 testability bus interfaces to nodes within board level UUTs to build a backplane level bus that also extends to other subsystems (backplanes) to build a system level bus. In Figure 10-28, the P1149 testability bus also extends beyond the board level to the system level. The smart card also makes use of the P1149 bus for on-card self-test. In Figure 10-29, the P1149 testability bus is used on-card by the smart card for on-card built-in test. The test manager card uses the P1149 bus to receive status data from the smart card while acting as a built-in test resource in its own right to test the dumb card.

P1149 T-BUS CONFIGURATION OPTION # 1

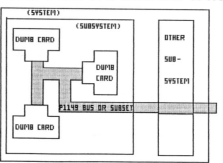

"DUMB CARD" HAS NO SOPHISTICATED BIT PROCESSOR ON IT
CARD INTERFACE IS A BACKPLANE OR OTHER CONNECTION SCHEME

FIGURE 10-27. P1149 bus configuration option 1. The testability bus, if dealing with all dumb cards, would typically be carried in its own format to other subsystems for control and evaluation of test data.

This is an example of a combination of centralized and distributed built-in test. The test manager card could also use the P1149 testability bus for its own on-card self-test.

In Figure 10-30, the test manager card gathers status data at the backplane level from both dumb and smart boards, uses the testability

P1149 T-BUS CONFIGURATION OPTION # 2

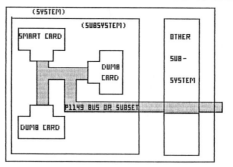

"SMART CARD" HAS BIT/BIST CIRCUITRY ON IT

FIGURE 10-28. P1149 bus configuration option 2. If a smart card is present, it could control and monitor unit under test nodal data via the testability bus in its own subsystem as well as data from other subsystems.

P1149 T-BUS CONFIGURATION OPTION # 3

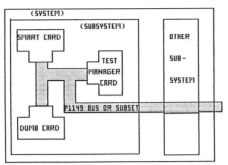

FIGURE 10-29. P1149 bus configuration option 3. A test manager card could be placed in one of the subsets to control information flow to and from both smart and dumb cards in its subsystem as well as other subsystems.

bus on-card for its own self-test purposes, and then converts the P1149 bus to another bus format altogether before transmitting testability data to other subsystems. The P1149 testability bus can be slaved to any other type of system level bus whenever a specific system design requires that option.

P1149 T-BUS CONFIGURATION OPTION # 4

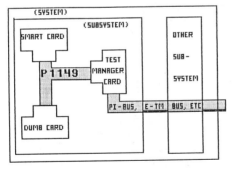

"DUMB CARD" HAS NO SOPHISTICATED BIT PROCESSOR ON IT
"SMART CARD" HAS BIT/BIST CIRCUITRY ON IT
CARD INTERFACE IS A BACKPLANE OR OTHER CONNECTION SCHEME

FIGURE 10-30.   P1149 bus configuration 4. A test manager card could convert P1149 testability bus data to some other format for transmission of control and visibility data, or the results of local diagnoses based on that data, to other subsystems.

TESTABILITY BUSSES AND ATE

The addition of testability busses to devices, boards, and systems does not unduly complicate the interface between the unit under test and the test system. It does make the ATE system much more efficient, both for go/no-go testing and especially for fault isolation. The following sections detail some of the intricacies of interfacing ATE to testability busses and show the advantages of it.

Introduction to Testability Bus Interfacing

Interfacing the testability circuitry to automatic test equipment is a straightforward task. The inputs and outputs from the circuits that come off the board and connect to the ATE (referred to as the testability bus) can be thought of as "windows" into the circuitry under test.

When operating the testability circuitry in the parallel (addressable real-time) mode, and when it is desired to control an internal node, for example, a stimulus vector from the ATE is output to the controllability circuit to load a 1 or a 0 into the address of the particular control point to be activated and the control point enable (TENA*) line is activated with a logic 0.

When it is desired to "look at" an internal visibility point, the visibility point address is placed on the testability bus. The state of the internal visibility point can then be seen at the visibility point data output. This data can be compared with a 1 or a 0 that is part of the test program's "expect data" (i.e., it is part of the good response vector).

Testing High-Speed Internal Signals

If the testability bus is implemented with a testability chip set consisting of control and visibility circuits which have combination serial and real-time addressable capabilities, it is possible to use the testability circuits to test fast internal nodes by letting the testability circuits act as latching buffers for those high-speed internal signals.

In the serial mode, serial test vectors can be input to the controllability devices in advance of their parallel application to the UUT, or even in between stimulus vectors to the primary inputs and outputs of the UUT (depending on the speed at which the tester and the UUT operate). On the visibility side, the latching capability of a testability chip set means that data can be captured from the UUT and stored in the visibility circuit until the ATE has time to evaluate it by scanning it out.

This capability means that even the fastest boards can be tested with today's ATE.

An example might best explain that statement. Suppose a board with 20-MHz internal signals must be tested on a 5-MHz tester. Data can be latched into the visibility circuit at test steps 1, 2, 3, 4, and so on, corresponding to the board's internal clock steps of 1, 5, 9, and 13. The data can be evaluated at a 5-MHz rate while the board's internal circuits continue to operate at 20 MHz. Test steps 5, 6, 7, and 8 would look at internal clock steps 2, 6, 10, and 14. Then test steps 9, 10, 11, and 12 would capture clock steps 3, 7, 11, and 15. Finally, test steps 13, 14, 15, and 16 would capture clock steps 4, 8, 12, and 16 (see Figure 10-31).

Thus even though the tester couldn't examine the 20-MHz visibility data in real time, it could examine it all by sampling every four steps and incrementing the sampling strobe time until all of the board's internal clock steps had been checked. The board under test, in the meantime, continues to run at its 20-MHz internal data rate.

The test programming is a little more complex to accomplish this sampling approach, but the testability circuitry allows fast boards to be tested on slow testers. Thus a little investment in extra test programming (the cost of which has been drastically reduced in the first place by adding visibility and control) can save a large new capital expenditure and extend the life of existing test equipment.

The examples in this chapter were presented to give you several ideas on how to implement the P1149 standard testability bus or combinations of its subsets in various electronic product designs. As mentioned earlier, they are by no means the only methods that can be used to implement the P1149 bus, and the user is encouraged to be creative in making the best designs, from a functional and performance standpoint and from a testability and maintainability standpoint.

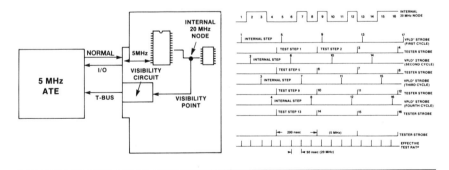

FIGURE 10-31. Testing high-speed internal signals. A sampling approach is needed if a slow test resource is to be able to test fast signals in a unit under test. Shift registers with parallel load capability can provide this interface to the testability bus.

11

Mechanical Guidelines

This chapter deals with designing products mechanically for factory and field test requirements. It also provides the mechanical engineer (product designer) with design-to-test guidelines that will assist in

- Developing a product that can be readily accessed and serviced and is fully testable at the component, module, and system levels
- Lowering the cost of developing special test equipment, adapters, and fixtures
- Avoiding unnecessary production test and field service time and cost

OVERALL TEST PHILOSOPHY

In the development of electronic equipment, it is evident that the product designer, in conjunction with the electrical designer and project manager, must implement mechanical testability in the early phases of the program to answer the following questions:

- Will the design of the unit facilitate fabrication, assembly, checkout, and field access and meet the testability goals for the product?
- Can existing ATE interface fixtures be utilized?
- Can the need for special tools, test setups, test procedures, and test equipment be minimized?

Since the mechanical design for testability of most units is unique to each application, a firm set of "do's" and "don'ts" is difficult to establish. To make the most use of design experience over the past years, this chapter lists design parameters and goals as well as the various techniques for implementing them.

ACCESSIBILITY

Accessibility for troubleshooting and repair is a key mechanical testability requirement. Accessibility guidelines fall into the following major categories:

- Human engineering
- Packaging
- Quarter-turn fasteners
- Fuses
- Slides for subassemblies
- Cables and service loops
- Visual indicators
- Test points/test connectors
- Radio-frequency interference (RFI) shielding

Human Engineering

Locate equipment controls, adjustments, and displays in functional groupings so that they can be operated and viewed from a single operator position. Consider maximum viewing distance, angle, and illumination limitations of a test setup. Consult with industrial design if necessary.

Factory testing may often be more detailed than field maintenance. Factory test and inspection points may end up in awkward locations when trying to keep field test points accessible. Try to keep factory test and inspection points and field test points accessible to cut down on troubleshooting time.

Packaging

The best design-to-test criteria for this case would be to use the fewest screws or fasteners that the specification allows. During the fault isolation process, particularly with large military electronic systems, a large portion of the repair time is spent removing and replacing covers.

Quarter-Turn Fasteners

Quarter-turn fasteners, per MIL-F-5591, can be used for nonstructural closure. Quick-operating, high-strength panel fasteners, per NAS-547 or MIL-F-22798, can be used for structural retention. Quarter-turn fasteners may also be used in commercial products. In either case, they make disassembly and reassembly faster and easier.

Fuses

Locate any fuses, or circuit breakers, so that they can be seen and replaced, or reset, without removing other parts or assemblies.

Drawer Slides

Place rack mounted units on slides provided with limit stops and cable retractors wherever possible. See Figure 11-1.

Drawer/Assembly Accessibility

Avoid the tendency to jam as much as possible into a box. Things don't always have to be small just for the sake of smallness. Think about it.

Cables and Service Loops

Design and route cables so that they are accessible and can be disassembled without unsoldering. Development of system and subassembly

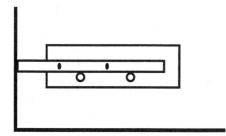

FIGURE 11-1.   Drawer slides for subassemblies. Drawer slides make it easy to extend a subassembly out from its rack for testing, troubleshooting, calibration, and repair.

wiring and cabling between the electronics is sometimes one of the last design tasks before release of the product to manufacturing and quite consistently creates major problems in the testing and maintenance of the hardware. Typical types of problems encountered are

- Cable obstruction of subassemblies and components
- Inadequate service loops for subassembly removal
- Wires pinched and damaged by doors, covers, and structure
- Improper cable routing in equipment
- Inadequate wire and cable identification
- Difficulty in cable removal and replacement

Cable Origins

Label or color-code each wire in a harness or cable so that it can be traced from origin to termination. This also will reduce troubleshooting time and the mean time to repair (MTTR).

Operational Test Points

During the design of any unit, it can be readily seen that certain points can be defined as the prime operational input or output points. These points should be in the most prominent place available for testing. See Figure 11-2.

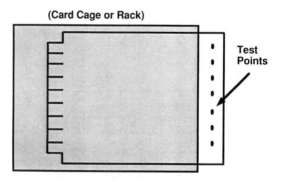

FIGURE 11-2.   Operational test points. Operational test points, particularly for critical or often used measurements, should be conveniently placed for access without placing boards on extender cards.

Visual Indicators

Can indicators be viewed without removing access covers? If externally visible LEDs can indicate which internal PCB needs replacement, the test operator may have the replacement part made available before the unit is disassembled, reducing the MTTR.

RFI Shielding

All access covers, whether hinged or detachable, should contain captivated, quick-operating fasteners that meet RFI requirements. If possible, provide hinged covers for access with transparent windows for viewing.

CONNECTORS

Connectors are the physical access between the circuit under test and the outside world. Their implementation and use are thus critical to the success of any electrical design for testability approach.

Test Points/Test Connectors

System functional and field maintenance testing should only require gaining electrical access to the UUT through its I/O pins and/or test connectors. All system level test points and adjustments should be brought out to facilitate testing at this level. Test and repair of such modules requires that the designer plan this process carefully and help to develop efficient test and repair procedures.

Ease of Disconnection

Be aware that much time is spent connecting and disconnecting plugs and receptacles. If specifications allow, use bayonet-type or similar disconnect types of connectors.

Color-Coded Tabs for Critical PCBs

To flag a safety hazard, color-coded PCB extractor tabs may be used. High-voltage PCBs would stand out if, for example, the tabs were red.

Reference to Mating Connectors

Code each connector plug to the receptacle to which it is being mated (e.g., P1 to J1, P2 to J2, etc., and not P1 to J2!).

PCB Extenders

Provide capability for extender cards so that both sides of a PCB are accessible for testing. See Figure 11-3.

Use of Empty Card Slots for Testing

Units containing PCB card mounting cages should also contain one or more extra card slots and an extender card for test diagnosis and field maintenance use.

Solderless Connections

Any field replaceable part or subassembly should be pluggable or quick-disconnectable.

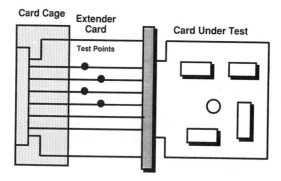

FIGURE 11-3.   PCB extender boards. When extender cards are used, it is often handy to provide "turrets" or some other connection scheme on the extender card itself for measurement instrument return lines.

Keyed Connector Receptacles

Physically similar but noninterchangeable parts or subassemblies should be keyed to prevent the possibility of inserting the wrong unit and damaging the equipment.

BOARD LAYOUT GUIDELINES

Substantial savings in test fixturing costs are available if certain board layout guidelines are followed during product design. Many of these guidelines take very little board space for implementation. They do, however, take forethought and discipline.

Standard PCB Layout

Common orientation of components is desirable for automatic insertion. Locate discrete components, DIPs, or flat packs on a given unit using a common orientation (not at 90 degrees or some odd angle to each other) so that automatic insertion, soldering, and termination techniques may be used. See Figure 11-4. A convention like this is also convenient for manual probing during troubleshooting, since time is not wasted trying to find the pin 1 indicator.

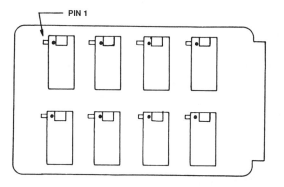

FIGURE 11-4.   Component orientation. When components are oriented the same way, it saves time and reduces errors for both assembly and test operations.

Extension Pads

Often an existing or standard test fixture pattern can be used to test standard sized printed circuit boards if test points are placed on a 0.100-inch (or 0.050-inch) grid, all at the same height, and have a minimum probe contact surface area of 0.035-inch diameter. Any component mounting hole, pad, or post that cannot be located on the grid can be brought onto the testable grid pattern by an extension pad.

Use of Standard Fixtures and Test Equipment

Review all testing requirements for application on existing factory test equipment and fixtures. Try to avoid designs which require the development of unique pieces of test equipment or special fixtures.

Registration Holes

Always provide a pair of 0.125-inch minimum diameter tooling registration holes in the product design item, except for very small submodules or ceramic substrates. These units should be fabricated, inspected, tested, and assembled by using a specifically identified set of three locating points on two adjacent surfaces of the item.

Ground Points

If the designated test equipment requires a UUT ground, it should be on a standoff that is accessible to the various types of clip leads in use. Frequently, short circuits have occurred because the clip lead has sprung off during tests.

Individual Leads

Care should be taken during design to eliminate, or at least minimize, the use of individual leads for test point access. The best design to test criteria would be to have all testing done through the edge connector (or other test connector).

Modular Functions/Feedback Loops

Partition the equipment into specific modular functions or subfunctions such that each module or subassembly can be functionally tested as an entity.

Solder Mask

The use of a solder mask is desirable when isolating test points or points that will require external jumpers and for reducing the number of solder shorts incurred during manufacturing of a PCB.

Components on One Side Only

Bed-of-nails test fixtures require that components be placed on only one side of the PCB. Any jumpers or engineering changes which require additional wire routing should always be placed on the component side of the assembly. This will help insure that the additional components and wires do not interfere with the ATE interface.

The only exception to this guideline is in the case of surface mount technology passive components (i.e., resistors and capacitors). These can be placed on the secondary side without much impact on fixturing as long as they are placed far enough from test pads to insure that no interference occurs.

Spacing between Components

Many ATE interface fixtures require the use of IC clips for nodal access. A minimum spacing of 5 mm will help facilitate the use of these devices. See Figure 11-5.

ADJUSTMENTS

Anytime a manual adjustment is required, the whole testing process has to be monitored by an operator who must intercede when required. If no adjustments were needed, the unit could be tested automatically, in groups, or overnight, and so on. Adjustments should be minimized.

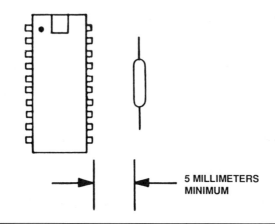

5 MILLIMETERS
MINIMUM

FIGURE 11-5.   Component spacing. Enough space should be left between components so that probes and IC clips can be used during testing and troubleshooting.

Minimum Nominal Selection

The mechanical and electronic engineers should get together to minimize the nominals (i.e., select-on-test parts). The best case is to eliminate the necessity completely. If nominals are the only way, keep the range of selection low. A large range together with a big production lot could have severe cost impact on spares that may never be used but must be available and priced.

Switch Guards

Sensitive control adjustments should be located and guarded so that they cannot be accidentally disturbed.

Conformal Coating

Conformal coated modules and potted subassemblies pose a particular test and maintenance problem. Such assemblies should be fully tested at the subassembly level before coating or potting. Set up test and inspections at assembly and subassembly levels before the coating or potting processes. Also make sure that enough testability connection access is left after coating to allow for efficient and economical functional testing of the assembly after it is coated (or potted).

Critical Measurements

Critical measurements requiring high-complexity, high-accuracy measurements should be avoided whenever possible. If the specification or test requirements demand measurements and tolerances that really push the state-of-the-art in test equipment capabilities, that fact should be highlighted in some form. Critical state-of-the-art measurements require much more time, exotic test equipment, and drive up the production cost. The factory could easily overlook critical test requirements if those requirements were buried within the test specification.

Be careful not to specify tests that reverify the design characteristics of the unit under test each time. The objective in manufacturing is to make sure that the (presumably fully characterized and verified) design was duplicated correctly. In the field, the objective is to return a once-working piece of equipment to full service.

OTHER PHYSICAL GUIDELINES

Several other mechanical (e.g., physical packaging) guidelines are worthy of note.

- Place logically equivalent faults in the same device package. A four-gate wired-OR circuit, for example, should be constructed using four gates contained in the same package, not four gates in four individual packages.
- Physically separate analog and digital assemblies.
- Provide a "short-circuit link" on printed circuit boards to allow ATE to verify the alignment between the board under test and the test connector.
- Keep the mechanical interface to the board under test as simple as possible.
- Place test pads close to the device pins to which they are connected.
- Make unused (from a functional standpoint) inputs on components accessible and controllable.
- Provide an adequate ground plane structure in order to suppress noise from high currents induced during either normal circuit operation or testing.

12

Surface Mount
Technology Guidelines

Another new technology that is causing testing difficulties, especially for in-circuit testers, is surface mount technology (SMT). Also sometimes called surface mounted components (SMCs) or surface mounted devices (SMDs), components in the SMT family are placed on the board rather than through the board.

Thus many critical visibility and control lines are no longer accessible to the fixture on the bottom side of the board, and some of the component package styles make it impossible to probe the component from the top side of the board. The lure of SMT is higher circuit complexity and higher package density—more performance in less space. SMT components have much closer lead spacings than traditional through-hole components as well, and this makes fixturing and probing more difficult.

As in almost every case where new technology is used in electronics design and manufacturing, there are powerful reasons for incorporating SMT in new products. The actual device packages and the packaging process for the semiconductor manufacturer can be much less expensive than the current plastic or ceramic dual in-line (DIP) packaging approach. And while SMT component prices may currently be slightly higher than DIP packages, that price difference will be reversed when high-volume production of SMT components becomes the norm. The real allure, however, of surface mount technology is simply more product performance in less product space.

But the wholesale move to surface mount technology is being slowed, longer than anyone predicted, by (among other things) the lack of knowledge and techniques for testing these new more complex boards. The device packages have leads that cannot be probed with bed-of-nails fixtures (see Figures 12-1A and 12-1B). Or, if they can be

FIGURE 12-1A.   No probe targets for J-lead parts. These J-leaded parts do not go through the board and cannot be probed from the top side of the board.

FIGURE 12-1B.   No probe targets for gull wing part. These gull wing part leads can be probed from the top, but that is not recommended because the parts tend to shift during the soldering operation.

probed, it takes very expensive so-called clam shell fixtures that can access the board in two or three axes (see Figure 12-2).

It is also expensive to rework SMT printed circuit boards. Replacing through-hole components is a relatively well understood, straightforward task—not so with surface mounted components. They take extra care and special tools for removal and replacement. And they are more expensive, particularly if they are application specific or custom components.

FIGURE 12-2.   Expensive fixtures required. Double-side or even three-dimensional fixtures can be fabricated to deal with surface mount technology, but they are expensive and not highly reliable.

The smaller SMT packages typically use the J-lead-type pin configuration. And the pin counts for surface mount packages, since pins can be placed on all four sides of the devices, are higher than they were for DIP packages.

There are more variations on leaded and leadless chip carriers for even higher-pin-count devices. Pin counts above 100 are now common and are steadily increasing as more and more functionality is placed on each semiconductor device.

Another popular surface mounted component packaging style is the mini-flat pack, which is also difficult to probe with a fixture. The 60-pin package has leads on 0.050-inch centers, while the smaller 40-pin package has leads on 0.040-inch centers. So while expensive fixtures can be used for the larger device, probing on centers with under 0.040-inch spacing is an art that has yet to be reliably and affordably perfected.

Thus new testability guidelines, both for in-circuit testing and for functional testing, have been developed to enable test engineers to write test programs and do fault isolation more efficiently. Especially with components mounted on both sides of the board, access for troubleshooting becomes very difficult and thus functional testability guidelines must be incorporated into SMT designs.

As is typical with SMT boards, while there is considerable circuitry contained on the board, the number of edge connector input/output lines may be very limited. This means that designers must make provision for bringing critical control and visibility points either to a special probing area (for the bed-of-nails fixture), to a test only input/output connector, or to the edge connector (either directly or using some sort of testability interface circuit).

There are new fixtures that can access the top of the board, the bottom of the board, and the edge connector pins of the board. Yes, it can be done, but at what cost? Not only is the initial fixture expensive (about $20,000), but the dedicated probe plates (three of them in this case) are much more expensive than standard dedicated bed-of-nails fixtures. And with smaller, more closely spaced probes (which typically have less travel or stroke than standard probes), contact reliability can become a significant problem.

MECHANICAL GUIDELINES FOR SMT BOARD DESIGN

There are some mechanical design solutions to the problems of accessing SMT PCBs with bed-of-nails fixtures. They deal mostly with making sure that the fixture can use probes spaced on 0.100-inch centers, limiting parts placement to one side of the board, providing test pads for components whose leads cannot be probed, and grouping tall components so that they do not interfere with the stroke limitations of the spring probes.

Board Size

The first SMT testability guideline is to keep the size of the printed circuit board under control reasonably small. As board size increases, the ability to probe it with a bed-of-nails fixture decreases. One thousand 7-ounce spring probes on a standard through-hole technology board of 8 by 10 inches will exert a force against the vacuum force of the fixture of 7,000 ounces, or about 437 pounds. With an area of 80 square inches, the vacuum force required to pull the board down is about 5.47 psi.

Converting the 8 by 10 inch board to SMT, with SMT's typical 4 to 1 increase in component density, would mean a total of 4,000 probes. This would raise the force required to 28,000 ounces (1,750 pounds) and would require a vacuum force of about 21.9 psi to pull the board down. This is no mean feat. Denser boards may end up being completely untestable by vacuum fixturing techniques, and larger boards may cause accuracy problems. In the event that a board is too dense for vacuum fixturing and is populated on both sides with active components, the only alternative for probing is to use a mechanically actuated fixture.

Components on One Side

The second mechanical guideline is to mount components only on the top side of the board. This guideline will continue to be violated on a regular basis, since one of the advantages of surface mount technology is its ability to mount ICs on both sides of the board! Putting the chip capacitors and resistors (and other small chip components such as diodes) on the bottom side of the board does not present much of a problem compared to putting ICs there. So even though keeping large components on only one side of the board is a significant SMT testability guideline for in-circuit testing, it is very difficult to enforce.

Test Pads

A third guideline for in-circuit testing (or bed-of-nails fixturing for functional board testing) is to use test pads to avoid probing the component leads themselves. Probing the component leads can mask bad solder joints (when the pressure of the spring probe causes enough of a connection for the device to pass the test) that will result in system failures (when the spring probe pressure is removed). Most ATE and fixturing vendors recommend that test pads be placed on 0.100-inch centers to facilitate probing.

Another risk with trying to probe the components themselves is that, because the components are not mounted rigidly through the board but can move during the soldering process, targeting problems can result. The same part, from board to board, may be skewed in any direction by as much as 0.010 inch and still make good connections to the etch on the board. But that skew may make a probe miss the target, or it may cause the probe to jam, bend, or break if it hits the device itself (or even the edge of the lead) instead of the device lead.

Tall Components

A fourth mechanical guideline is to keep test pads away from tall components. The so-called short-stroke probes used for accessing test pads on 0.050-inch (or 0.025-inch) centers (and below) cannot be used to access test pads on the board under test if the test pad is placed next to a tall component. There is simply not enough plunger movement available.

Some probe vendors have introduced long-stroke (0.250-inch plunger movement) probes for 0.050-inch center probing. But the previous guideline still holds. If test points are placed next to a 1-inch high heat sink, no probe is going to reach them. Using daughterboard probe plates is, of course, another possible "patch"solution, but long, unsupported structures typical of fixture daughterboards are weak and will not provide satisfactory probe life or targeting accuracy.

Some people recommend staggering test pads to provide access via standard 0.100-inch grid spring probes (see Figure 12-3). With this method, the test pads for alternating pins are placed at different distances from the component leads to be accessed. In Figure 12-3, 0.050-inch center component leads are staggered out for access by 0.100-inch center fixturing. The only problem with this approach is that it takes considerable additional board space, negating some of the space/performance advantages of SMT. If you are using dedicated, rather than on-grid, fixtures, the space penalty is not as large because you don't really need the full 0.100-inch clearance in both directions. The trade-off is in the price for one standard universal grid fixture versus several dedicated off-grid fixtures.

Another approach to spreading out test pads for access with standard fixtures is test pad *fan-out* (or *spidering*). With this technique, enough room is left around each surface mounted component so that even close-centered leads (such as 0.025-inch centers) can be fanned out to provide 0.100-inch center probing targets. See Figure 12-4.

Here again, depending on the component lead spacing and the area required to fan out the test pads, considerable board space may be

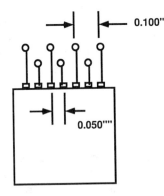

FIGURE 12-3.   Staggered test pads. To make it easier to use standard fixtures with surface mount components, test pads can be staggered as shown. Pads should be far enough from component leads so that component drift will not interfere with probe access.

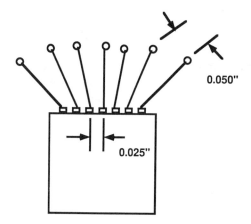

0.050"

0.025"

FIGURE 12-4.   Test pad fan-out. An alternative to staggering test pads is to fan them out as shown. This is sometimes the only alternative with parts with very closely spaced leads.

needed to render the board testable with conventional fixturing techniques, even if probing every electrical node does not require probing every SMT component lead. If one is dedicated to bed-of-nails fixturing techniques, however, fanning out test points may be the only way to make the board testable. Whether test pads are staggered or fanned out, there are two critical guidelines for the test pads themselves.

Test pads should have a minimum area equivalent to a pad with a diameter of 0.035 inch, although larger is better. Anything smaller does not present a probe target that can accurately and repeatably be hit. Test pad spacing is also important. Test pads should be placed at least 0.025 inch from component leads to insure that, if the component is skewed during the soldering process, the probe will still not hit the component instead of the test pad.

ELECTRICAL GUIDELINES FOR SMT BOARD DESIGN

There are also some electrical solutions to making surface mount technology boards more testable (for both in-circuit testing techniques and functional testing techniques, but most importantly for functional testing). These electrical techniques include multiplexed test points, on-board diagnostics, and built-in test.

Many forward-looking electronics manufacturers are beginning to dedicate anywhere from 5 to 10 percent of board hardware and software real estate, just as semiconductor device manufacturers dedicate up to

20 percent of silicon space for testability and built-in test. They realize that a small increase in board size, or putting fewer functional components on each board and adding an extra board if necessary, while seemingly more expensive when looked at only from the parts cost standpoint, results in major product cost savings when all of the factors (parts cost, assembly labor, test equipment, test programming, test fixturing, testing, troubleshooting, and rework) in the business process are considered.

Test Point Multiplexing

Test points can be multiplexed (see Figure 12-5) so that multiple functional circuit internal nodes can be brought out to a limited number of edge connector (or test only connector) pins or probing points. You can use a 1-of-8 decoder whose inputs are connected to critical functional circuitry visibility points. Test point addresses are supplied (again from edge connector or test only I/O pins) to select which test point signal will be output from the decoder.

Access to these critical internal nodes significantly reduces the time it takes to write a test program and the number of guided probing steps required to isolate faults on a functional tester. It does increase parts cost, but that is more than offset by the lower testing costs.

Control on SMT Boards

Extra control of SMT board devices can also considerably enhance the ability to perform accurate fault isolation on a board. Accurate fault isolation is even more critical with SMT components than it is with

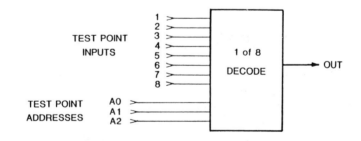

FIGURE 12-5.   Multiplexed test points. A multiplexer can be used to bring a larger number of test points to a smaller number of physical pins at a card edge connector or test connector.

through-hole parts since it is more expensive to rework an SMT board than it is to rework a standard technology board.

Many functional tester guided probing systems can diagnose faults beyond the node *if* they have access to chip enable or tristate lines. Some extra inverters and pull-up resistors (see Figure 12-6) can provide individual control over the devices whose outputs are bussed together. The on-board diagnostic control lines are brought to an edge connector (or test only connector) so that chips can be individually disabled and any changes in the voltage or current on the faulty node measured to determine which chip on the bus is actually at fault.

SMT Board Built-in Test

It is also quite realistic to provide a separate built-in test processor right on the SMT board, as illustrated in Figure 12-7. With this method, test point data are stored in ROM for access by the built-in test processor. Some sort of test point interface device is installed between the built-in test processor and the critical nodes of the functional circuitry.

The built-in test processor can access any test point and cause it to be compared with the "good behavior" data stored in ROM. Any failures can then be indicated (either physically via an indicator or by sending a message back to the built-in test processor for further action). The standard testability bus can often be used when implementing circuitry such as that shown in Figure 12-7.

The total long-term solution to testing surface mount technology boards is to render them testable during product design, to make sure that testability takes advantage of the power present in most processor-

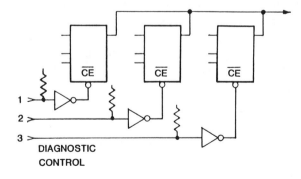

FIGURE 12-6.   Inverters for tristate control. On bus-oriented boards, it is important to have control of all of the devices on the bus. This makes it easier to determine which device is causing a bus line fault.

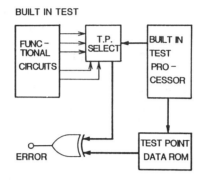

FIGURE 12-7. Built-in test for SMT boards. SMT boards designed with testable functional circuits are usually quite good candidates for built-in test. Known good circuit data can even be sotred in on-board memory.

based boards to drastically lower test programming costs, and to make SMT boards testable on functional automatic test equipment.

Built-in self-test is becoming an economic necessity. Those electronics manufacturers who intend to compete in world markets will include it. Those that don't include both inherent testability and built-in test in SMT board designs will find that their future ability to compete is literally at risk.

Visibility and control of internal nodes from functional ATE, both to reduce programming costs and to aid in fault isolation, must also be designed into SMT boards. This built-in control and visibility can be thought of as distributing a limited bed-of-nails fixture right on the SMT board and bringing the testability bus (which accesses multiple internal nodes via a few edge connector or test only I/O pins) out to the functional test equipment interface. The extra visibility points, in particular, reduce the need for operator probing during diagnostics, since the ATE system can probe internal nodes simply by addressing each built-in visibility point. Remember, every visibility point built into a board design has the potential for reducing troubleshooting costs by 50 percent or more per fault.

13

Software Guidelines

This chapter deals with considerations that should be applied when software is being designed. Many of these concepts are applicable to both system and subassembly tests. Many also require interaction between system level designers and hardware and software engineers. Thus close cooperation and teamwork is necessary for software to be successfully testable.

HARDWARE DESIGN FACTORS REQUIRED FOR SOFTWARE TESTABILITY

Hardware design engineers must provide, in the system, subsystem, and module designs, sufficient circuitry to accomplish BIT/BIST/BITE requirements. The exact requirements will vary with each system design and should be specified in the system specifications. These hardware requirements include ROM space allocation, BIT/BIST/BITE test points, and the means for routing BIT/BIST/BITE data, as well as the test data, so that the test software can accomplish the required testing.

GENERAL SOFTWARE DESIGN GUIDELINES

It is very important that diagnostic aids be included with the operating software program. Here are several of the general guidelines to be followed for enhancing the testability of your software.

Operator Controls

The initialization of the system setup by the test operator should be provided for, as should provision for individual test selection, initialization, and recording of test results (for analysis and historical purposes).

Operator Options

Operator options that should be provided include

- Halting the test on occurrence of an error and display of the last successfully executed step
- Continuing the test, despite errors, with critical variable parameters displayed to the operator. Single-stepping should also be included as part of this basic capability
- Looping on a failed test, with error reporting as a selectable option
- Isolating faults
- Looping on a "sequence of tests" capability
- Displaying all errors to the video terminal or printer
- Collecting historical data during execution of test such as
  - Pass/fail indications only
  - Identification of failed parts
  - Number of times the test passes or fails
  - Actual and expected results with appropriate identification

SPECIFIC GUIDELINES FOR TEST CONTROL

The test software should provide for initialization, control, and execution of those tests defined in the BIT/BIST/BITE disclosures for the UUT. The test software control modules, as a minimum, should include

- Test initiation and termination at the test operator's request
- Test repetition for a selected number of repetitions
- Test sequence definition and initiation
- Repetition of the test sequence definition and initiation as stated previously
- Test option processing
- Test failure history compilation
- Manual operator control over all of the foregoing test sequences
- Operator control of individual test options

SPECIFIC GUIDELINES FOR TEST MODULES

Each diagnostic test module should provide

- The ability to output a hard copy of the test stimulus data required for execution of all tests defined for the UUT
- Test evaluation tools, including operator assistance displays, if required

Standard test module options should include

- An output of actual versus expected results
- A "halt on error" option
- Fault isolation diagnostic information to the test operator
- The capability to output all error messages
- The capability to output all operator messages
- Operator assistance to output all options and the test selection menu (HELP menus)

SPECIFIC GUIDELINES FOR SYSTEM LEVEL DIAGNOSTICS

When built-in test circuitry is included, there are several guidelines for its successful programming.

BIT Circuitry and Test Points

The effectiveness of a diagnostic program is highly dependent on the types and quantity of BIT/BIST/BITE circuitry and on the quantity and strategic placement of monitored test points within the equipment. These must be planned for and provided by the equipment design engineers early in the design phase.

Stimulus and Response

Stimulus and response data for each diagnostic test must be defined at the system/subsystem level. The same data should be planned for use in factory testing and in field maintenance of the equipment. This keeps the amount of unique software to be written to a minimum.

Manual Control

Provide for manual control of test sequences so that each test can be selected individually and appropriate test combinations can be executed at the operator's discretion.

Fault Isolation

The diagnostic test for all subsystem (unit) testing should assess the unit's operability and isolate failures to replaceable items. Some basic guidelines for fault isolation considerations of diagnostic tests include

- Designing each test so that it will
  - Execute independent of all other tests
  - Diagnose to a functional portion of the unit
  - Initiate upon successful completion of higher-priority preceding tests for this unit
- Designing fault isolation routines so that the results of only one independent test have to be analyzed. If fault isolation requires analysis of the results, the last test in a multitest sequence should analyze all of the results
- Having each independent unit test provide both a go/no-go status indication and fault isolation to the cause of the failure
- Wherever possible, making each test capable of being terminated prior to completion and of being reinitiated at its start point (either automatically or at the option of the operator)
- Designing all software so that it is structured by test priority. The test software should take advantage of both subroutine constructs for all message outputs and of failure dictionaries which identify the location of the most likely failed replaceable unit

Response Conditions

Circuits should be designed into the units under test so that they can accommodate the following subsystem response modes:

- Incorrect response from the subsystem, including "no response" conditions
- Inconsistent response content conditions

- Unexpected response conditions
- Incorrect response content conditions

Hard-Core Tests

Insure that the program includes a bootstrap, or equivalent, function which will establish that a minimum working instruction set (MWIS) and memory are working. This MWIS will be used to establish that other instructions are working, always using proven instructions, until the entire instruction set has been verified.

MEMORY TESTS

Memory tests are then executed to validate that all associated memories are working. Some representative memory test patterns are discussed in this section.

Zeros

Writing 0's sequentially at each address in memory, the test system (or BIT/BIST/BITE software) then reads the addresses sequentially. This simple N-type test quickly examines either cell opens or cell shorts and the ability to store 0's. However, its main use involves verifying the operation of the hardware interfaces to memory elements.

Ones

The system writes 1's sequentially at each address in memory and then reads the addresses sequentially. Except for checking the ability to store 1's instead of 0's, the Ones pattern serves the same purpose as the Zeros pattern.

March

After writing a background of 0's to memory, the system reads the data at the first address and writes a 1 to this address. The same two-step read/write procedure continues at each sequential cell until the system reaches the end of memory. Each cell is then tested and changed back to

0 in reverse order until the system returns to the first address. Finally, the test is repeated using complemented data (i.e., writing a background of 1's to memory). An N-type pattern, the March pattern, can find cell opens and shorts, as well as address uniqueness faults and some cell interaction faults.

Galloping Pattern (Galpat)

Into a background of 0's, the first cell (test cell) is complemented and then read alternately with every other cell in memory. This sequence continues as every memory cell eventually becomes the test cell. The system then executes the same sequence using complemented data. With an execution time proportional to the square of the cell count (N-squared type), Galpat looks for open cell opens, cell shorts, address uniqueness, sense amp interaction, and access time problems (especially those faults due to the address decoder delays).

Column Disturb

Into a column of 0's, the system writes complemented data continually (for a specified time) to the first and last cells in one column and then reads data in all other cells in that column. Next, the system restores 0's in the first and last cells and disturbs the second and second-to-last cells, after which the data in the first and last cells are read. This sequence continues for each column in memory. The system then repeats the entire sequence using the complemented data. Designed to find disturb sensitivities and refresh sensitivities in dynamic RAMs, the pattern's execution time depends on the number of disturb cycles executed.

Block Ping Pong

The address sequencing remains identical to that of Galpat, but the background data consists of alternating blocks of 1's and 0's, and users can determine the block length. While the Block Ping Pong pattern has similar execution time and fault-finding capabilities to Galpat, it can also locate some data sensitivity problems that remain undetected by Galpat.

Surround Disturb

Into a background of 0's, the system complements the first cell (test cell) and repetitively reads the eight physically adjacent cells (up to 255 times). After reading and restoring the test cell to 0, the system continues this procedure until each memory cell has been the test cell. Then the sequence is repeated for complemented data. Surround Disturb finds possible adjacent cell disturb malfunctions. Execution time varies with the number of disturb cycles executed at each test cell.

Write Recovery

The system writes the second cell to a 1 and reads the first (test cell) to a background of 0's. Then the second cell is restored to 0, and the first cell is read again. The same read/write sequence repeats between the third and the next test cell, the fourth and the next test cell, and continues to the last cell and the last test cell. The entire sequence is repeated until every cell has acted as the test cell. Finally, the system repeats the pattern using complemented data. An N-squared test, the Write Recovery pattern, primarily locates write recovery–type faults, although it can also find faults listed under Galpat.

Walking Pattern

Into a background of 0's, the system complements the first cell and reads all other cells sequentially. After reading and restoring the first cell to 0, the system complements the next cell and reads all other cells sequentially. This procedure continues for all memory cells. The pattern is then repeated using the complemented data. The N-squared–type Walking Pattern examines memory devices for cell opens and shorts and address uniqueness.

Sliding Bit

Not in itself a pattern, Sliding Bit merely generates a shifting data pattern and then repeatedly calls a test such as March or Galpat to check data bit uniqueness in multi-data-bit chips or boards.

Checkerboard Read/Write

The Checkerboard Read/Write program writes a data checkerboard (alternating 0's and 1's) into memory. A control program or delay subroutine then executes a delay before the system reads the checkerboard pattern. NOT-Checkerboard Read/Write patterns can provide complemented data pattern. Usually employed in conjunction with a long delay between the write and read parts of the test, the N-type checkerboard patterns (with delay) evaluate static performance and data retention in static RAMs.

Other Memory Test Patterns

Other memory test patterns include Address Test, Moving Inversion (MOVI), Row Disturb, Row Galpat, Column Galpat, Sliding Diagonal, Buffer Row Galpat, Buffer Column Galpat, Buffer Adjacent Galpat, Buffer Write Surround Disturb, and Buffer Ping Pong. The actual selection of the appropriate test pattern(s) depends on the system design architecture and the test requirements for the system. The objectives are maximum fault coverage and highest system reliability.

SPECIFIC GUIDELINES FOR LRU TESTING

- The software resident in the system and the BIT/BIST/BITE should provide for testing of the lowest replaceable unit (LRU) and all its interfaces. It should include fault detection capabilities in the interface tests.
- Normally, the fault detection and isolation test procedures are produced by the cognizant design engineering activity in the form of diagnostic flowcharts and are accompanied by a narrative description of the flowcharts. The flowcharts and narrative description are called BIT/BIST/BITE disclosures and are given to software design engineers for implementation.

TEST SOFTWARE DEVELOPMENT PLANS

A test software development plan is based upon the results of a system analysis. This plan defines the test software requirements and the design approach for the software. Structured programming principles should be followed as a strict rule. Remember to keep subroutines in a separate program area, with single entry and single exit points for each subroutine. This will ease both the development and the testing of software.

System Analysis

An analysis of the following types of system requirements for each program is required to determine test software requirements and design approach. Items to be considered from an overall system standpoint include

- Built-in test
- Built-in self-test
- Built-in test equipment
- Factory test plan
- Throwaway modules versus repair
- Maintenance levels (depot, organizational)
- Spares philosophy
- Mean time to repair
- Environmental requirements
- How systems are deployed
- Test equipment concept
- Special customer considerations
- Test software concept

System analysis from a software point of view will encompass

- System test and fault isolation overview
- System requirements (BIT/BIST/BITE)
- Test and troubleshooting aids
- Test equipment recommendations/considerations/selections
- Fault isolation criteria (system, unit, PCB, etc.)
- Software requirements
- Computer sizing and selection
- Training level of user personnel

14

Testability Documentation

This chapter deals with the design and scope of the documentation supporting a PCB, unit, test set, or test software program. Although it may seem only a secondary consideration to many, well-designed documentation can greatly improve the testability of a particular UUT or program.

TEST SOFTWARE DOCUMENTATION

The purpose and objective of test software documentation is to inform factory test personnel of how to use and maintain the delivered software to perform the test functions associated with the UUT. A program written in assembly language normally demands more documentation than one written in a higher-level language (e.g., C, ADA, ATLAS). All test programs should provide for adequate test software maintenance and user information. Two types of basic documentation are required to support test software: a user's manual and maintenance manual.

User's Manual

The user's manual is designed for use by test operators, and describes how to use the delivered software with the unit under test. It should provide for test selection by the operator and include relevant information on use and operation of the tester. The manual should include

- All step-by-step operating instructions necessary to set up and initialize the tester and the UUT for testing

- A list of all test hardware and software required to perform the test
- A description of the program loading and test execution procedures and all test options (looping, isolation, etc.) provided for by the test software

Maintenance Manual

The maintenance manual helps maintenance personnel troubleshoot the test software. The manual should reflect the final logic in the software program after system test. The manual should include

- A program listing in the language used (C, ADA, ATLAS, etc.) with complete comments. The comments are considered complete if they permit a "qualified" reader with no prior knowledge of the program to understand the program without having to refer to the program statements. The body of comments should represent a complete functional description of program statements, grouped by function
- A gross level flowchart and a narrative description of the control software which describes sequencing/execution, I/O processing, and interrupt handling
- For each software module, a detailed flowchart with narrative description of the test rationale and data flow should be provided. This flowchart should
  - State the test objective
  - Describe the test software (not hardware) module, including
    - program modules (identification and definition of program tasks and major subsystems)
    - module interface definition (description of intermodule communication and how modules are interrelated)
    - storage required for program and data
    - data organization (i.e., stand-alone, top-down, separate module for use with another program, certain resources assumed or other assumptions, etc.), including the identification of all stimulus and response data
  - Give the program origin—whether the program is derived from or is a modification of another program. List any standard software packages used
  - Describe the information flow between test control and the individual tests. This description should contain the name

> calling sequence
> inputs
> outputs
> modifies (or destroys): (e.g., registers, global variables, hardware flags, etc.)
> calls (e.g., subroutines, other programs, etc.)
> description
> other information as appropriate
> entry point
> exit point
> from (i.e., was the program control transferred from this module?).

- Give the test program confidence. List the percentage of possible faults found by this program and the methodology used to obtain the measure of confidence (e.g. simulation, hardware fault insertion, etc.). Also list the faults not detectable by the entire test program. These faults will have to be identified at later testing steps

HARDWARE DOCUMENTATION

The purpose of the hardware documentation is to provide both factory and field test personnel with all material relevant to the operation and troubleshooting of the hardware assembly. The information should be structured in such a way as to make troubleshooting as easy as possible. The following are items of hardware documentation:

- Schematic diagram
- Relevant waveforms, timing diagrams, and logic diagrams
- Wiring diagrams and wiring run lists
- Assembly drawings and parts lists
- Copies of manufacturer's specification sheets for all components contained on the unit under test
- UUT functional description and theory of operation
- Voltage/resistance chart for UUT nodes
- List of test equipment required
- Equipment performance specifications and test requirements
- Test flow
- Block diagram
- Brief description of trade-offs (reasons for decisions)
- Faults found at each test level (include method of measurement)

- Interface
- Graphic description of interface
- Nodal cross-reference (bed-of-nails)

Each documentation package should be as complete as possible, although not all UUTs will require the same level of documentation. The complexity of the UUT and the design and type of tester used will both affect the documentation required. For example, a digital PCB tested on a logic tester using signature analysis will require a nodal signature table for all PCB nodes and a good block diagram and theory of operation. An analog PCB, tested manually, will require waveforms for appropriate nodes and a more detailed (component level) theory of operation.

Schematic Diagrams

There are several points to keep in mind in the design and layout of the schematic diagram, which can greatly improve the testability of the UUT.

- All input pins should be shown on the left of the schematic, and all output pins should be shown on the right. I/O pins should not be shown in the middle.
- If possible, nodal signatures, relevant waveforms, and/or timing diagrams should be included on the schematic at the appropriate nodes. If not, the schematic should reference the document(s) containing this information. It would also help to number the nodes on the schematic and to reference the node number on the document containing the waveforms (see Figure 14-1).
- Functional designations (CK, R/W, Q, BUSRQ, etc.) should be shown next to each IC pin number on the schematic, except on logic gates. Logic gates should have the input signal names listed at the inputs, and the signal name logically formed at the output.
- Power supply circuits on non-power-supply PCBs should be shown in a single location on the schematic, and all voltages should be labeled.
- Schematics of all subcircuits, such as vendor-supplied modules mounted on the PCB, should be provided on the overall

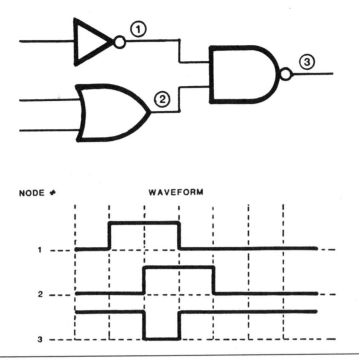

FIGURE 14-1.   Node-referenced timing diagram. Knowing which signals appear at each node, and when, helps when it comes to generating test vectors and doing fault isolation.

schematic or supplied separately and referenced on the overall schematic.

- The schematic should reference the assembly drawing and give the part number of the next higher assembly.
- If the schematic for a single PCB or assembly takes up more than one page, all interpage signals should be referenced with signal name and/or number and should show the zone designation and page number of where the signal goes to and comes from.
- Do not show a single I/O pin more than once on a schematic without cross-referencing the zone designations.
- Do not use more than one logic symbol to depict a specific component or hardware part. If a vendor-supplied module includes a schematic with symbols which are different, redraw the vendor's schematic or correct those symbols which are different. Different representations can be very confusing.

Relevant Waveforms, Timing Diagrams, and Logic Diagrams

As mentioned earlier, waveforms and diagrams are best shown on the schematic where space permits. When shown separately, they should be referenced to the schematic by schematic name and/or number and node number. In addition:

- All voltage levels should be shown.
- If timing is important, state where an oscilloscope, logic analyzer, or other measuring device should be triggered.
- Show all necessary specifications and tolerances (such as pulse width, rise and fall times, etc.).
- When depicting a logic diagram for a given IC type (such as a J-K flip-flop), reference IC numbers on the schematic.

Wiring Diagrams and Wiring Run Lists

Point-to-point wiring diagrams for all wiring harnesses, as well as wiring run lists for all wire-wrap boards, should be provided. These should include a list of points wired together, color and size of the wire, signal name, and, in the case of a wire-wrap board, the level at which the wire is wrapped to a pin. A technician can more easily trace incorrectly connected wires or shorts from a wiring diagram or wiring run than from a schematic diagram.

Assembly Drawings and Parts Lists

The assembly drawing and related parts list should be as complete and as simple to read as possible. Avoid overcrowding an assembly drawing with unnecessary details. Information on the parts list should include specific information such as resistor tolerances and capacitor working voltages. This is particularly helpful when parts must be substituted due to shortages, unavailability, and the like.

UUT Functional Description and Theory of Operation

The UUT Functional Description and Theory of Operation should begin with a brief description of the function of the UUT being tested (i.e., memory PCB, power supply, D/A converter, etc.) and how it fits into the overall unit or system.

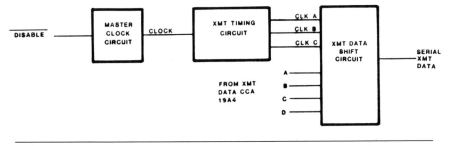

FIGURE 14-2. Example block diagram. A block diagram gives a good overview of the overall functional blocks of a system.

The theory of operation should contain a block diagram and a description of all functional sections of the UUT. For instance, if a group of ICs forms an oscillator, the ICs can be described as such. It would also help to label these ICs as "oscillator" or "master clock circuit," or whatever, on the schematic surrounding the ICs with a dotted line if necessary for clarity. Include a separate block diagram with the description (see Figure 14-2).

A detailed theory of operation may or may not follow the block diagram, depending on the detail and type of troubleshooting required. If the UUT is digital and is being tested on a digital logic tester using a guided probe for fault isolation, a detailed theory of operation might be nice but not necessary, since the technician will be following tester-provided probing directions rather than actually analyzing the functions of specific ICs.

If, on the other hand, the technician is really expected to get into the circuits, node by node, with an oscilloscope or other piece of diagnostic equipment, a detailed theory of operation could be vital. Remember that field maintenance personnel do not normally have sophisticated ATE at their immediate disposal.

Generally, a theory of operation should always be provided. Also, analog circuits are almost always candidates for a detailed theory of operation, because the schematic alone may not give the technician sufficient information due to the variety and complexity of most analog circuits.

Voltage/Resistance Charts for UUT Nodes

A voltage/resistance chart is generally useless for digital circuits but may be helpful in analog circuits, particularly power supplies. Each node on the UUT has a resistance-to-signal (as opposed to chassis) ground when the UUT is off, and each node has a voltage level when the UUT is on. The voltage/resistance chart can supply this information to aid in troubleshooting. If this type of chart is supplied, be sure to specify the type of meter being used.

List of Test Equipment Required

All equipment required to test the UUT should be listed, and a drawing of the suggested test setup should also be supplied. Important factors that should be noted along with the list are part numbers of connectors required in the test setup to interface with the UUT, types of coaxial cables required, and terminations or loads required in the test setup.

UUT Performance Specifications and Test Specifications

Performance specifications and UUT I/O tolerances should be listed. When choosing tolerances, be careful not to make them so tight that more UUTs fail than pass a test. Make a special note of unusual or abnormally tight tolerances for the technician. Test specifications should be clear and easy to understand.

A step-by-step test is often the easiest type of format to follow. An alternative, especially for more complicated procedures, is a flowchart with subroutines to aid in isolating faults. No matter what type of test procedure format you choose, always provide information on what the operator or technician should do if a step or/and entire test fails. It is difficult to isolate a fault when the technician does not even know what signal or function he or she is checking at a particular test step.

15

Implementation Guidelines

This chapter outlines a methodology for implementing testability into a design or project program and provides a model policy and procedure for making sure that testability, among other things, is considered during design reviews. Your operation may already have a formal program in place. If it does, the model in this chapter may give you some suggestions for improving it. If it does not, the model that follows should give you at least a starting point in setting up your own program.

TESTABILITY PROGRAM FLOW

Testability is most cost-effective when incorporated into the design at the initial product conception. To retrofit testability after the product has been released to manufacturing can be very expensive and make a modification for testability unjustified due to the cost of implementation. Shown in Figure 15-1 are the various stages of product development where testability can be added to a design.

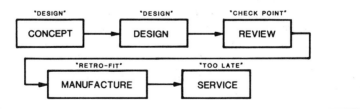

FIGURE 15-1. Testability implementation stages. It is seldom cost effective to "retrofit" for testability. It may still be too expensive and time consuming to "checkpoint" for it and correct deficiencies by redesign. To design it in, both at the concept and detail levels is very inexpensive, very quick, and pays a large return on investment.

One of the major tasks in any testability program is to implement its concept into the normal design procedures of the organization. The flow diagram in Figure 15-2 accomplishes that objective. There may be various versions of the flow, depending on the amount of testability required and on individual program constraints, but the program in Figure 15-2 is a good place to start.

The testability program can use existing policies and procedures to implement its concepts and to provide the checks and balances in the design process. The key to the success of any design, along with the testability aspects, lies in the design review. The concept of testability utilizes the design review methodology as its cornerstone to insure compliance with testability considerations.

DESIGN REVIEWS

Design reviews for the verification of functional conformance to operating specifications are normally held by most organizations. Missing sometimes, however, is the testability aspect of the design review. Many times it is often helpful to hold a specific design review just for testability. The following subsections outline the purpose of doing so and the steps to be taken, not just for testability but for many other design aspects that may also be often overlooked.

Purpose of Design Reviews

The whole idea behind the design review for testability is to catch potential test problems in the initial stages of the design. This helps to eliminate many engineering change notices which otherwise might have to be implemented. A very important part of any effort to improve design for testability is a design review program. Most corporations have such a program. However, they usually do not include a review of the testability of that design by a test engineer. We feel that it is necessary to have the design reviewed by many people, each looking for the elements relating to their areas of expertise. A serious testability problem may easily be overlooked by someone primarily concerned with the functionality of the board. For this reason, test engineers are encouraged to attend the design reviews.

Design Review Policy

The following sample design review policy is offered as a model upon which a specific policy, tailored to the individual organization, can be based.

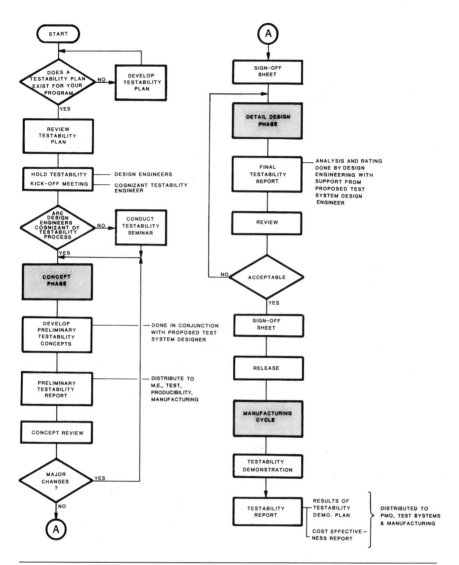

FIGURE 15-2. Testability program flow. This flow is being successfully used by many organizations around the world. It is a good model from which to build a customized flow for a specific organization.

Purpose. An organization's products should be designed to meet standards of performance, manufacture, reliability, ease of maintenance, and cost goals. Design reviews are intended to provide a systematic appraisal of the engineering concepts and detailed design approach, measured against appropriate specifications, objectives, and standards.

Responsibility

1. Division managers are responsible for insuring compliance with this policy.
2. Program managers (product manager may be substituted for program manager throughout this policy whenever there is no program manager) or their delegates are responsible for establishing, scheduling, determining the personnel complement of, acting as chairpersons of each design review on each piece of equipment to be reviewed, and for insuring the distribution of the appropriate preliminary information to all reviewers. A published policy delegating the determination of the personnel complement of, and the chair of, each design review to another clearly specified individual will be considered as being in agreement with this policy.
3. Engineering and manufacturing managers are responsible for insuring the attendance of their representatives at design reviews.
4. The responsible design engineer will present the design for review. All pertinent and necessary technical information, including requirements, limitations, goals, present design, major design decisions, and alternative solutions considered on any unsolved problems will be supplied. He or she is responsible for the design of the product and its conformance to specifications.
5. Each member of the review team is responsible for evaluating and constructively criticizing the design.

Definitions. Design reviews are defined as periodic, planned, and systematic appraisals of design concepts, alternatives, and objectives and are accomplished by knowledgeable specialists (a) before and during the engineering phase of a program or project and (b) during the engineering phase of a design change for articles that have already been released to production.

Design Review Procedure

This section presents a model design review procedure upon which a specifically tailored procedure can be developed for any organization.

General. Design reviews provide the opportunity for the most knowledgeable engineering and manufacturing specialists to offer constructive criticism of the design in order to

- Insure that contract and specification requirements and their implications are fully understood
- Insure contract compliance
- Preclude obvious omissions or errors in compliance to either specifications or good engineering or manufacturing practices
- Anticipate and eliminate, if possible, problems that may reasonably be expected to occur
- Lead to a timely design that can be economically manufactured
- Evaluate the design against design goals
- Provide the plan for redesign, if necessary

Design reviews require adequate preparation of all participants (design, manufacturing, and support) before the meetings.

Frequency of Design Reviews.   All phases of design, from the proposal through production, are to be covered. Five design reviews are recommended as follows:

1. A preproposal review, or in the case of projects having a concept design phase, the equivalent of a preproposal review
2. A concept review
3. An electrical review
4. A mechanical review
5. A release review

Concept, electrical, and mechanical reviews should be mandatory. Preproposal reviews, or their equivalent, should be mandatory unless specifically waived by the division manager. Release reviews are mandatory if electrical or mechanical reviews have resulted in major changes.

Design Review Team.   A team is required to review each design. Selection of the review team is made by the program manager, or his or her delegate, after consultation with the responsible system and design engineers. In addition to the system and design engineers, plus one or more representatives from manufacturing, a team will include the following specialist capabilities as appropriate:

Circuit design	Mechanical design or packaging
Reliability	Materials application
Maintainability	Test equipment design
Component application	Human engineering
Heat transfer	Quality control
Environmental	Purchasing

Value engineering   Advanced development or research
Producibility   Safety

Scheduling Design Review Meetings.   The program manager, or delegate, will determine which design reviews are required and will schedule them at a time mutually satisfactory to the responsible design engineer. Each participant must be given adequate information in time to prepare for the meeting (usually at least 5 to 10 days).

Conduct of Design Review Meetings.   The responsible design engineer (senior systems engineer in the preproposal review) presents the design to the review team in a manner that permits critical analysis. Included should be

- A brief description of the requirements of the particular unit or units being reviewed
- The functional relationships and interfaces among all units providing an input to, and accepting an output from, the unit or units being reviewed
- Specific difficulties, either encountered or anticipated, in complying with the specific requirements
- The proposed alternatives for eliminating either existing or anticipated problems
- The manufacturing processes to be used
- At electrical, mechanical, and release reviews, a prediction of the product cost in the required or estimated production quantities, compared with the cost target

When the information presented to the team is not sufficient to permit critical analysis, the design review chairman should postpone the review, point out the areas of deficiency, and reschedule the meeting. Each member of the design review team, in the design review meeting, offers specific, constructive criticisms of the design. Adequate preparation is essential to the accomplishment of this task.

The design review chairperson is also responsible for insuring that a permanent record of the proceedings of the meeting is maintained. Minutes published by the design review chairperson listing all recommendations with their acceptance, rejection, or requirement for further study should be sent to the responsible design engineer's supervisor, the manager of engineering, and to those present at the design review.

At preproposal and concept reviews, the formal contract and specification requirements for the complete system and for the various elements being reviewed must be clearly set forth by the senior systems engineer or a representative from the program office.

DIGITAL T-SCORE RATING SYSTEM AND CHECKLISTS

Several years ago the U.S. Government spent a considerable sum of money coming up with a rating system for digital printed circuit boards. The objective was to provide a means to actually measure, via calculations before hardware was committed, the testability figure of merit of a digital printed circuit board design.

The original work has been considerably refined over the years. The digital T-score rating system in Appendix B is the refined version, and the testability checklists in Appendix A can be used to help make sure that the proper amount of testability has been included in the design.

If a design gets a low score with the T-score rating system, it will be difficult to write test programs for it and to test, troubleshoot, and service it. If a design does not answer yes to most of the checklist questions, a similar situation is most likely to exist.

16

Test Techniques
and Strategies

This chapter describes the various testing strategies currently available for board level assemblies. The basic capabilities and limitations of each piece of inspection level and functional level test equipment are summarized, along with an overview of the test flow during production of a given product.

PRODUCTION TEST FLOWS

Each subassembly (PCB, cabinet, wiring, etc.) is normally tested prior to final assembly. After final assembly, the unit is tested alone and again as part of a system.

The objective of a testing strategy is to remove faults as early in the manufacturing process as possible. This is because the further along in the process that a fault must be isolated and repaired, the more expensive is the operation. Finding a faulty component by testing the component will normally cost hundreds of times less than finding the same faulty component once it has been installed in a system or subsystem.

Not all manufacturing operations will use all the available options. Which ones are chosen will depend on the types and quantities of faults estimated to occur during the manufacture of a given product or product line. Other factors to be considered include the volume of products to be tested and the variety of different components, boards, and systems.

CABLE, BACKPLANE, AND BARE BOARD
CONTINUITY TESTING

Several manufacturers provide equipment that is used primarily for cabinet/interconnect wiring continuity testing. Most operate automatically and at high speed, and there are a number of available options. Major features and capabilities of this type of equipment, depending on vendor and options chosen, are

- High-speed testing of point-to-point wiring in cables, harnesses, wire-wrap panels, and multilayer PCBs
- High-pot for detecting transient insulation breakdown to 1.5 kV
- Resistance measurements from 0.01 ohm to 1000 megohms, using a voltage stimulus of 5 to 1500 VDC, with ±5 percent accuracy
- DC voltage measurements from 0.1 mVDC to 1500 VDC with ±1 percent accuracy
- Adjustable delay time and dwell for high-stability measurements
- Programmable parameters so that the UUT can be tested on a go/no-go basis and so that any embedded resistors can be measured

LOADED BOARD OPENS AND SHORTS TESTING

Opens and shorts testing for loaded boards is a testing philosophy based on the statistics that most PCB defects are historically due to problems related to opens and shorts. Most opens and shorts are a result of the assembly and solder stages of the manufacturing operation.

Performing opens and shorts testing on loaded boards increases productivity and lowers costs by off-loading simple faults from a more complex ATE system. The fixturing is via a bed-of-nails, and it is typical to fixture the assembly to detect only shorts at this stage in the manufacturing process. This is because, for through-hole board designs, 90 percent of the faults are typically short circuits. To fixture a PCB for both opens and shorts testing requires a test nail at the ends of every trace and is more costly than simple shorts testing, which requires only one nail per node.

For surface mount technology boards, opens are the more common failure mechanism. Only one connection per node, however, is required for testing open circuits for surface mount technology boards as long as all nodes are brought to the bottom side of the board and can be probed. The open circuit will be detected when the manufacturing defects test is run on each component. Depending on the level of sophistication of the tester, it may or may not be able to detect opens related to SMT digital ICs and may only detect opens related to the passive components. It may still be necessary to use in-circuit or functional testing to detect open circuits associated with digital SMT ICs.

The objective of loaded board opens and shorts testing is to isolate simple faults (e.g., those points that are connected and should not be, and those points that are not connected and should be).

This testing technique differs from bare board testing in that fixturing is typically connected to one side of the board only, and this technique picks up manufacturing process-induced shorts and opens that were not present at the bare board test level.

Benefits
- Isolates process-induced shorts and opens at an early stage
- Requires small programming effort
- Low operator skill requirements

Limitations
- Little or no parametric capability
- No functional test capability

IN-CIRCUIT INSPECTION BOARD TESTING

In-circuit testing is the technique of testing each component on a printed wiring board on an individual basis, without regard to its intended circuit function. This testing philosophy is based on the following assumptions:

- If all components are of the correct value, and they are all correctly installed, the PCB will function correctly.
- The majority of defects on PCBs are process-induced.

In-circuit testers are very good at detecting and isolating manufacturing-induced faults on a PCB. They test digital ICs for their basic truth tables and are able to make resistance, capacitance, and inductance measurements.

Electrical guarding is used during the power-off testing of discrete components on an assembled PCB. Guarding helps to reduce the effect of other components on the device under test (DUT). Due to the circuit configuration, it is often very difficult to get a good measurement due to parallel components and "sneak" paths to ground. There are some component configurations that cannot be effectively guarded. For example, parallel bypass capacitors on a power bus can be impossible to test with diagnosis to the failing component, especially if there are 20 or so of these devices on a bus and the fault is that one or more of the capacitors are open.

The bed-of-nails fixture should contact each node. There is little need for making contact at each end of every trace, except on critical bus lines.

Benefits
- Onepass diagnostics
- Easy to program
- Easy board handling
- Low operator skill requirements

Limitations
- Some untestable components in-circuit
- Tolerance of tester versus tolerance of UUT
- Component interaction not checked
- Limited functional test capability

Typical stimulus and measurement parameters for this type of equipment include:

- Programmable short detection from 5 to 20 ohms.
- Resistance measurement range from 1 ohm to 9.99 megohms, with an accuracy of ±1 percent from 1 to 10 ohms, and an accuracy of ±0.5 percent at greater than 10 ohms.
- Capacitance measurement range of 10 to 999 pF, with an accuracy of ±1 percent from 10 to 1000 pF, and ±2 percent at greater than 1000 pF.
- Inductance measurement range of 1 mH to 10 H, with an accuracy of ±3 percent from 10 mH to 10 H, and ±5 percent at less than 10 mH.
- Capability of supplying a constant current of 10 nA to 99.9 mA with an accuracy of ±5 percent.
- Capability of supplying a constant voltage of 0.1 to 99.9 V with an accuracy of ±0.5 percent.
- DC voltage measurement range of 0.01 to 99.9 V with an accuracy of ±0.5 percent.
- Capability of in-circuit differential voltage measurements in a powered circuit without altering circuit performance. This can be used to measure DC voltages in the presence of an AC or DC common mode voltage level. The measurement range is 0.1 to 15 VDC with a common mode voltage of ±15 V_{p-p} maximum with an accuracy of ±0.5 percent.

MANUFACTURING DEFECTS TESTING

There is also a class of testers called *manufacturing defects testers* (MDTs) or *manufacturing defects analyzers* (MDAs). These testers are basically either beefed-up opens and shorts testers or stripped-down in-circuit testers. They have been developed because full-capability in-circuit testers have tended to become more expensive and have, in some cases, migrated to combination in-circuit and functional capability systems. The manufacturing defects testers typically perform opens and shorts testing and tests for missing, wrong, and wrongly inserted components as well as resistance, capacitance, and inductance measurements. They typically do not check digital device truth tables because this is normally the realm of the full in-circuit tester.

A manufacturing defects tester can many times replace an opens and shorts tester and an in-circuit tester and provide almost the equivalent fault coverage of the in-circuit tester for manufacturing-type defects only. Since the manufacturing defects tester does not exercise the functions of the digital or analog components on the board under test, some faults typically found on an in-circuit tester will not be found on the manufacturing defects testers and thus must be found at a later (typically functional) test stage.

DIGITAL FUNCTIONAL TESTING

There are several types of digital functional testing that are in reasonably wide use today. This section outlines the philosophy, methods, technology, benefits, and limitations of each technique.

Static Functional Testing

Static functional testing involves testing a PCB as a functional entity rather than testing each component individually. It involves sequencing the test vectors at rates lower than the normal operating speed of the UUT. This technique assumes that if the assembly functions properly, then all components must be correctly sized and installed and that process-induced problems can't exist.

Fixturing of the UUT to the ATE is usually via the edge connector. Some static functional testers employ a limited bed-of-nails fixture for increased visibility to the UUT.

Input stimulus vectors are provided to sensitize faulty circuit paths and to exercise functions that the board will perform in the next assembly. If a fault occurs, the ATE prompts the operator with a sequence of node names that allows him or her to probe from the failing output along a bad circuit path until the fault is isolated. Some static functional ATE systems use a fault dictionary for fault isolation. With this technique, the operator is provided with a fault reference number to a list of known faults for that assembly. The operator then matches the fault reference number from the tester to the number of the fault description in the dictionary. The dictionary lists all possible sources of potential error for any given fault reference number. A separate dictionary must be generated for each board.

The speed of test execution is usually limited by computer I/O rate (digital) for a go/no go analysis. For diagnostics, the limiting factor becomes the fault isolation time and number of operator probing operations required. For analog testing, the ATE is usually limited by the instrument conversion rate on the general-purpose instrumentation bus (GPIB). All tests are done at discrete stimulus/response setups, and this technique only requires a medium skill level for operation.

Benefits
- Verifies the functional integrity of the assembly
- Usually provides higher next assembly yields than previous methods

Limitations
- Does not test for timing-related faults
- Has a slow test execution with long test patterns
- Finds only one fault per pass

Dynamic Functional Testing

Functional testing, as described previously, involves testing assembled boards as a functional entity rather than as individual components. Dynamic functional testing involves sequencing the test vectors at the equivalent (or higher) operating speed of the unit under test. Dynamic testing is usually used for the type of board which contains ROM, RAM, a microprocessor, and various peripheral chips.

Speeds in excess of 20 MHz are usually referred to as dynamic at the board test level based on ATE manufacturers' definitions. Real dynamic testing usually means tests are performed at actual normal operating speeds of the unit under test. The difference between the two definitions is one of technology and terminology. Some devices cease to operate at speeds below about 800 kHz, and dynamic testing is imperative. In addition, high test rates can reduce production test times. "Soft" failures, such as pattern sensitivities, are more likely to be found with dynamic testing, regardless of whether it is ATE dynamic or real dynamic.

The internal architecture is similar to that of a static functional tester except that memory behind each pin is used to store patterns for rapid broadside stimulus and response vectors.

Signature analysis may be used to reduce large amounts of data to a single four- to eight-digit number and to reduce the large amounts of response data from the unit under test which must be analyzed by the ATE.

Bed-of-nails fixtures are not normally used due to high capacitive loading and noise considerations. New fixtures are being introduced which work around these problems, as new testers are introduced with high-speed functional and in-circuit capabilities on the same machine.

This type of ATE is becoming more popular due to the difficulty of finding the new types of soft failures and the need to approach true dynamic testing conditions. There are, however, certain limitations on the ATE, since good design practice means using mature (i.e., slow) components in the design of an ATE system, while engineers are designing new products to be tested using very high speed state-of-the-art technology.

Benefits
* Verifies functional and speed-related integrity of assembly
* Executes quickly

Limitations
* High cost of equipment
* Finds only one fault per pass

In-Circuit Emulation

In-circuit emulation is particularly useful in development applications. It is sometimes used for production and field service testing and requires good software design and, usually, a high level of skill for diagnostics.

This technique emulates the end-item operation via functional stimulus programs which are input to the board from the microprocessor socket. Loopback board design can be employed to provide a good test of peripheral and random logic. The technique also uses signature analysis to gather and analyze test data where the loopback technique cannot be implemented.

The emulator provides a "captive" microprocessor of the same family as used on the assembly, running at full speed, in a personality module. The PCB under test can run its own programs or special diagnostic routines stored in ROM. Test programs can then be structured for good repeatability of signatures. In-circuit emulation has the advantage of allowing the transportation of engineering programs to the testing functions provided proper care has been taken. The capital cost is relatively low, but usage requires careful design attention.

Stimulus speeds to 40 MHz and above are possible. The maximum speed for response data collection and analysis is currently about 25 MHz. Connection of the tester to the UUT requires a socket for the microprocessor. An RS-232C serial I/O port can interface with host computers. Large development systems are usually available with this type of tester for added flexibility in program generation and editing.

Benefits
- Provides for full-speed operation
- Verifies functional and speed integrity
- Can use existing self-test programs
- Can be used in field service test

Limitations
- Minimal diagnostic capabilities
- Microprocessor must be removed during test
- Requires high skill level in production test

Dynamic Reference Testing

Dynamic reference testing is a technique whereby input stimulus is provided to the UUT and to a known good board simultaneously. The outputs from the two boards are exclusive-ORed and monitored for failures. This type of ATE usually has automatic pattern generators as well as the ability to use stored patterns generated by the user. Unlimited response vector lengths are theoretically possible because the reference board acts like an infinite-length ROM for storage of diagnostic nodal data. Pattern speeds to 50 MHz are attainable with this type of ATE, and signature analysis techniques are used to verify that the known good board still functions properly.

Benefits
- Large numbers of test vectors can be applied at high speeds
- No LSI modeling required
- Verifies functional integrity

Limitations

* Requires manual test program generation
* Requires reference board for diagnostics

Signature Analysis

Signature analysis is not really a stand-alone test strategy, unless the input stimulus vectors are provided from a source external to the unit under test. Signature analysis is often used as an adjunct to several testing methods, including in-circuit and functional, and serves basically as a data compression technique.

Data compression is achieved in the signature analyzer by probing a logic test node from which data are input for each and every circuit clock cycle that occurs within a circuit-controlled time window. Within the signature analyzer is a 16-bit feedback shift register into which the data are entered in either their true or complement logic state, according to previous data-dependent register feedback conditions. In all, there are 65,536 possible states to which the register can be set during a measurement window.

These states are then encoded and displayed on four hexadecimal indicators and become a *signature*. This signature is then a characteristic number representing time-dependent logic activity during a specified measurement interval for a particular circuit node. Any change in the behavior of this node will produce a different signature, indicating a possible circuit malfunction. A single logic state change on a node is all that is required to produce a meaningful signature. Because of the compression algorithm chosen, measurement intervals exceeding 65,536 clock cycles will still produce valid repeatable signatures.

Serial data are shifted into the register along with start, stop, and clock signals. The remainder uniquely defines nodal states and times as long as enough patterns have been circulated through the shift register. Input stimulus vectors can either be provided by on-board software or from an external source such as an ATE system or in-circuit emulator.

Benefits

* Many thousands of tests can be applied at high speeds
* Fast program generation in many cases
* Large amounts of response data can be compressed
* Can be used for field service

Limitations
- Requires careful consideration in the design for testability
- Diagnostic resolution is poor in feedback loops and bus-structured boards

Dedicated Testers

A dedicated tester is typically a piece of in-house built test equipment designed for the expressed purpose of testing one type of circuit board. These testers can be designed to test both digital and analog boards. The diagnostic capabilities are typically not as good as those developed by the ATE manufacturers. The ATE manufacturers have each invested millions of dollars into the research and development of ATE.

These testers are usually designed in-house by the test engineering department and may not have good documentation, procedures, and so forth. They can, however, be tailored to a specific application, product, facility, or process. Dedicated testers are sometimes used when volumes are very high and many units are needed for throughput capabilities. Alternatively, dedicated testers are sometimes used when volumes are too low to justify commercial general-purpose ATE.

Benefits
- Tests can be tailored to the specific task
- Can be inexpensive if designed properly

Limitations
- Limited diagnostic capabilities
- High skill levels are typically required
- Can be very expensive if designed improperly

Hot Mock-up Testers

Another testing method used quite often is hot mock-ups. This technique is a form of a dedicated tester in that it is usually an in-house design and is dedicated to testing one type of circuit board. With this technique, a test system is configured using bits and pieces of the final product. All of the components of the test system are known good, except the unit under test. The system exercises the unit under test and monitors responses.

Hot mock-ups can be useful for QA audit purposes and test program improvement, providing that good records are kept. The capital equipment cost is usually buried due to the fact that inventory surplus can be used to build the tester. The technique is highly labor intensive, in most cases, due to limited diagnostics. However, it does provide the ultimate in functional testing.

Benefits
* Requires little design effort
* Inexpensive in terms of capital costs

Limitations
* Little or no diagnostic capability
* Requires high skill level operators
* Usually has long test times

ANALOG PCB TEST EQUIPMENT

Analog test equipment is typically specifically configured for use within a company. An example of such a specially designed piece of test equipment, described here, might be a system which is an automatic analog and dynamic logic test station. It is used to test PCBs and assemblies which are either completely analog or combined analog and digital. It is composed basically of a variety of separate pieces of commercial test equipment, a UUT interface panel, and a computer with disk memory.

The stimulus and measurement test equipment are controlled by the computer. Since the test should be entirely automatic, the PCB under test should not have extensive adjustments made to it during the test. The UUT interface panel, which typically contains such interface circuitry as loads, special stimulus circuits, and interconnections, must be designed specifically for each PCB under test. The various stimulus and measurement test equipment might contain

* DC voltage sources
* Signal sources
* Digital stimulus
* Switching matrix
* Digital measurement
* Bus programmable DMM measurements
* Bus programmable counter measurements
* Bus programmable waveform analyzer

It is also possible to configure manually operated test stations for such things as RF/IF circuit testing. Such a system would be used to test, align, and troubleshoot (to the component level) RF and/or IF assemblies. It would operate at frequencies up to 100 MHz and be composed of a variety of separate pieces of commercial test equipment and an interface panel. The interface panel might contain a variety of switchable attenuators, RF amplifiers, RF power dividers, mixers, and so forth, and could be used with all UUTs. The types of measurements that a station like this is typically capable of making include

- Gain
- Bandwidth
- Phase
- Standing wave ratio
- Insertion loss
- Swept frequency response
- Pulse characteristics
- Noise figure

Specially designed automatic test equipment may be installed and used to test, align, and troubleshoot both active and passive microwave assemblies, components, and networks. It is composed of various separate pieces of commercial test equipment, and a computer section capable of controlling all phases of testing for both stimulus and measurement test equipment. The types of measurements this kind of equipment is normally capable of making include

- VSWR
- Insertion loss (dB)
- Phase (transmission angle degrees)
- Gain
- Phase deviation (degrees)
- Loss deviation (dB)
- Flatness deviation from mean (dB)
- Group delay (nsec)
- Isolation (dB)
- Reflection magnitude
- Return angle (degrees)
- Return loss (dB)
- Transmission (real)
- Transmission (imaginary)

- Reflection (real)
- Reflection (imaginary)
- Z magnitude (mmhos)
- Z angle (degrees)
- Y magnitude (mmhos)
- R (ohms)
- X (ohms)
- R/Z_0
- X/Z_0
- G (mmhos)
- B (mmhos)
- G/Y_0
- B/Y_0

COMBINATIONAL TESTERS

Combinational testers are usually high-speed, high-performance test systems that include both in-circuit and functional (or performance, as it is sometimes called) test capability. They utilize a bed-of-nails fixture to overcome lack of inherent testability features in the unit under test and to brute force test patterns through individual ICs on the board (or clusters of components when they cannot be isolated for individual testing due to lack of testability).

The advantage of a combinational tester is that it can perform both manufacturing defects/in-circuit testing and functional testing of the board under test with only one handling operation for the board. Because it has access to every node (or at least most of the nodes) on the board under test, it also can overcome most common testability problems.

The main limitations of combinational testers are their high initial cost and their long and expensive test programming times and costs.

CHOOSING A TEST STRATEGY

The selection of the best test strategy in any given situation does not have to be, and indeed should not be, a guessing game or a matter of opinion. There is a relatively rigorous method for selecting the correct strategy in any given situation. That method is the subject of this section.

First, information is needed in order to develop a test strategy that will further the overall strategy of a business. Thus the test strategy developer needs information on the following criteria in order to formulate a good strategy:

- *Design criteria:* What technology will be used? (This may limit the number of choices or available testing options.)
- *Marketing criteria:* How many items will be sold? (This has crucial impact on the ability to pay for automated versus manual testing.)
- *Quality criteria:* How good does the product have to be, and, therefore, how good does the testing have to be?
- *Support criteria:* What product service strategy will be employed; if board swap is chosen, how will boards returned from the field be handled?
- *Application criteria:* Will the product be in a protected

environment or a severe environment? (This has impact on environmental stress screening options.)
- *Financial criteria:* How much money is available to finance the test strategy?

Data can be gathered from past experience, from consultants, from magazines, at trade shows, and even from competitors or companies who build noncompeting products of similar types and complexity as those contemplated for production in your facility. All early estimates should be documented so that they can be referred to in case they require revision.

At the incoming inspection stage, the decisions are relatively straightforward. If incoming failure rates are high (e.g., over a few hundred parts per million), incoming inspection is probably indicated. Simply calculate the cost to find defective components at board level and multiply by the number of faults that would be caused by defective components of each type. If it costs less to find them at board level than by implementing incoming inspection, skip the incoming inspection step (and vice versa).

At the board level, where diagnostics and rework enter the picture, things get a little more complicated. First, ascertain or estimate the board level fault spectrum that is anticipated. Figure 16-1 shows a typical fault spectrum for boards manufactured in the United States.

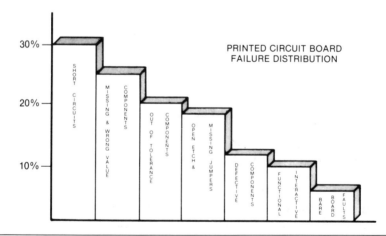

FIGURE 16-1. Typical printed circuit board fault spectrum. The relative frequency of occurrence for through-hole printed circuit boards is illustrated. The objective of a test strategy is to remove them as early in the process as possible.

The objective of the test strategy is to detect defects at the earliest possible step in the process, where they are typically least expensive to detect (and correct if necessary). Not all faults can be found, however, at every level. Bare board faults can be found at incoming inspection, by virtually any board test technique or even at the final system level. Functional interactive faults, on the other hand, can only be found at board or system level.

The major elements in the test strategy calculations include (at each level of testing considered)

- Capital equipment cost
- Test programming cost
- Test fixturing cost
- Test operation cost
- Diagnostic and rework cost
- Fault coverage impact on next level of test

At board level the last two are the most important on an ongoing basis. An in-circuit tester diagnoses multiple faults in one pass across the tester. A functional tester typically finds only one fault at a time, so, depending on the number of faults estimated per board, diagnostic time will have a relatively large impact on the testing strategy. Figure 16-2 illustrates the effect of multiple faults on costs using both strategies. One interesting thing to note about Figure 16-2 is the cost of diagnosing faults, using either method, when there are no faults. That's right, it's zero!

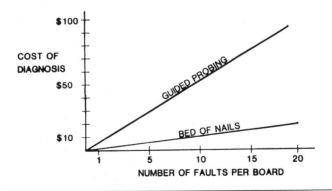

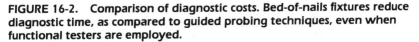

FIGURE 16-2.   Comparison of diagnostic costs. Bed-of-nails fixtures reduce diagnostic time, as compared to guided probing techniques, even when functional testers are employed.

How do you calculate the cost of each possible test strategy? The answer is relatively simple. Make a flowchart of each possible step in the strategy, create a formula for that flowchart, and then string the formulas for the various steps together to get a formula for a total strategy.

Take functional board test as an example. To calculate the operating cost of this step, construct a flowchart like the one shown in Figure 16-3. Once the flowchart has been constructed, create a formula that correctly calculates the cost of the operation. For the flowchart in the figure, the formula is

$$TC = M[(HT + TT) + (DT + RT + HT + TT)(F1) + (DT + RT + HT + TT)(F2) + \ldots]$$

Where TC = total cost, M = \$/hour, HT = handling time, TT = go/no-go test time, DT = diagnostic time (per fault), RT = rework time, and F1 and F2 are the portions of defective boards the first and second time through the process, respectively.

Anything that can be done to bring down the value of F1 will have a major impact on overall operating costs. That means defect prevention in the manufacturing process. Further, the next big factor is DT, the diagnostic time. Anything that reduces it (e.g., design for testability) will lower costs significantly. Handling and test times are usually the last elements you should try to improve.

A similar flowchart and formula for system level testing should then be constructed (including engineering evaluation, if required, of boards that pass the board test step but continue to fail at the system test step). The F1 and F2 rates at system test will be determined by the fault coverage (or test effectiveness) at the board level. F1 will be lower at system level for high-speed functional board test than it will be if only manufacturing defects testing is performed on boards prior to their assembly into the system.

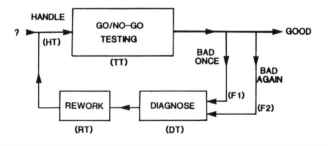

FIGURE 16-3.   Flowchart of test operation. Each operation can be flowcharted and its costs determined (or estimated). Large factors, such as F1 and DT, should be looked at first before looking at such things as HT.

Test coverage may thus be defined as

Total coverage = (percent of defects detectable in a design)
          × (percent of defects detectable by test equipment)
          × Percent of defects detected by test software)

Consider the following examples. In the first, use a design with 100 percent detectable faults and a manufacturing defects tester that can detect a maximum of 60 percent of the faults that could occur. Program it such that 90 percent of the faults that it can detect actually get detected. The fault coverage is then 54 percent. Thus 46 percent of the boards going into the system will have faults on them. In a second, consider a functional tester that can detect 100 percent of the faults, a board design with 95 percent detectable faults, and a test program with 90 percent fault coverage. The total coverage is then (100 × 95 × 90) percent = 85.5 percent, and only 14.5 percent of the boards going into the system will have defects.

Thus the board test coverage has a big impact on the system level F1 figure. Considering that diagnosis is typically 10 times more expensive at system level than at board level, the better job we do at board level, the lower our system testing costs. That is why all of the formulas for the various stages of a strategy must be added together, and the overall cost for a strategy must be examined as an entity.

Thus we would construct a formula for each strategy as follows:

Total cost = incoming inspection cost + board test cost
          + unit test cost + system test cost.

Cost, as used in this formula, includes capital equipment cost, test programming costs, test fixturing costs, and testing and troubleshooting costs on an ongoing basis.

Some examples will illustrate the use of this approach and point out the importance of looking at the whole process. The data base for the first example includes a system composed of 10 printed circuit boards, each made up of 100 components. A total of 1,000 systems per year will be produced, the manufacturing process yields 2.5 faults per board (one component fault and 1.5 workmanship defects), and the system yield is 85 percent after boards are tested on an in-circuit tester (with 95 percent system yield after functional board test).

The major cost elements are added and compared in Table 16-1. It is quite clear that the right strategy in this case is in-circuit board test. It is roughly half the price of the functional test strategy.

What happens, however, if we change the scenario? Raise the num-

TABLE 16-1.   Test Strategy Trade-off Example 1

Cost	In-circuit ($)	Functional ($)
Capital Equipment	250,000	500,000
Test Fixturing	35,000	5,000
Test Programming	30,000	80,000
Board Level Test	25,000	125,000
System Level Test	37,500	12,500
Total	377,500	722,500

ber of systems to be built each year from 1,000 to 5,000. Increase the complexity of the systems from 10 boards each to 50 boards each, which raises the cost of system level testing and diagnostics. Hold all other data the same. The results of this scenario are shown in Table 16-2. Functional testing is now clearly the strategy of choice. Why is that? If one looked only at the cost of the board test operation, ignoring the impact of board test fault coverage on system level testing and diagnostic costs, an incorrect choice would be made. Ignoring system level costs would result in the selection of the in-circuit test strategy, which is a half million dollar mistake because system test costs are very real.

A difference of only 10 percent fault coverage at board level in this instance makes a 3 to 1 difference in system level testing costs. This example illustrates the critical need to consider the entire manufacturing process when selecting test strategies rather than looking at each step in isolation.

The same formula and flowchart approach can be used to play "what if." What if I could raise board level yield from 70 percent (the average in the United States today) to 95 percent (the average in Japan today)? It may or may not be possible, but we can calculate the financial impact. Lower the F1 number from 0.30 to 0.05 and recalculate the test cost. The difference is the amount of money available to try to get the yield up from 70 percent to 95 percent.

TABLE 16-2.   Test Strategy Trade-off Example # 2

Cost	In-circuit ($)	Functional ($)
Capital Equipment	250,000	500,000
Test Fixturing	175,000	25,000
Test Programming	150,000	400,000
Board Level Test	125,000	500,000
System Level Test	1,875,000	625,000
Total	2,575,000	2,050,000

Perhaps that means new production equipment. Perhaps it means better assembler training. Perhaps it means better quality components coming in. It may not even be possible. But by running the calculation, you will know how much money you have to work with in trying to achieve your new goals.

The flowchart and formula approach is incredibly powerful. You can calculate the cost of testing with and without testability features. The difference in test costs is the amount of money available for added parts costs to implement the testability.

You can "what if" yields, fault distributions, fault coverage figures, and production quantities to see whether changes in conditions will have a major impact on your test strategy recommendations and decisions. You can find out what it is costing you to test, and then you can lower those costs.

Appendix A:

Testability Checklists

This appendix is a quick reference checklist that can be used before and during design reviews to make sure that no significant opportunities for testability improvement have been overlooked. It is divided into several sections to facilitate quick identification of the category of guidelines that may be of particular concern in each application.

These checklists contain most of the questions asked in U.S. MIL-STD-2165 (26 January 1985 issue) as well as additional questions relevant to the additional material throughout this text.

SYSTEM GUIDELINES CHECKLIST

Removable BITE Concept

Has the system been designed with BITE provisions?

Standardized I/O Pin Configurations

Have standard I/O pins been designated?

Minimum Use of Connector Types

Have the minimum number of connector types been used?

Extender Cables and/or Extenders

Have provisions been made to allow extenders to be utilized during test?

Test Points/Test Connectors

Have all critical nodes been routed to test points or a test connector?

Visual Indicators

Have provisions been made for visual status indicators during system test and diagnostics?

Ground Points

Have ground points been made easily accessible by an instrumentation ground clip?

Minimize Variable, Adjustable, and Nominal Selection-on-Test Conditions

Are there any components that need to be selected during test? If so, try to reduce the number of them to zero.

Digital Feedback Loops

Are all system level digital feedback loops controllable by the test equipment?

Generic Part Numbers

Are all components listed by their generic part numbers?

Component Reference Designators

Have reference designators been assigned to each component for quick location?

Is each hardware component clearly labeled?

Timing Diagrams

Have all critical timing diagrams been supplied with the system documentation?

Functional Packaging

Has each subassembly been designed as a functionally complete entity?

If more than one function has been placed on a board, can each be tested independently?

Within a function, can complex digital and analog circuitry be tested independently?

Within a function, is the size of each block of circuitry to be tested small enough for economical fault detection and isolation?

Are elements which are included in an ambiguity group placed in the same package?

Critical Measurements or Adjustments

Flag all critical measurements.

DIGITAL GUIDELINES CHECKLIST

Initialization

Have all sequential circuits been made initializable (e.g., MASTER CLEAR line included or an initialization sequence of less than 16 input clock cycles)?

Monostable Multivibrators (One-Shots)

Have the inputs to the multivibrator been made controllable?

Has the output of the multivibrator been made accessible?

Has the output of the multivibrator been made replaceable by the ATE?

Interfaces

Can the ATE use only one logic level to interface to the UUT?

Built-in Test Equipment (BITE)

Have provisions been made for BITE at the board level?

Feedback Loops

Have all critical feedback loops been made controllable by the ATE?

Oscillators/Clocks

Can all oscillators be controlled by or synchronized to by the ATE?

High Fan-in and Fan-out Nodes

Are test control points included at those nodes which have high fan-in?
Are test access points placed at those nodes which have high fan-out?

Bussed Logic

Have all system busses been brought to edge connectors so that there are no "internal only" busses?

Buffers

Have buffers been included for interface to the ATE where needed?

Visibility

Have test points been included for easy test equipment access?
Are unused (from a functional standpoint) edge connector pins used to provide additional internal node data to the tester?

Partitioning

Have logic functions been partitioned into logically separable units?
Are unused (from a functional standpoint) edge connector pins

used to provide test stimulus and control from the tester to UUT internal nodes?

Wired OR/AND Functions

Have wired logic functions been eliminated wherever possible?

Counters/Shift Registers

Have all counter chains been broken into smaller segments in a test mode, with each segment controllable by the ATE?

Redundant Circuitry

Are redundant elements in the design capable of being tested independently?

Programmable Devices

Are all programmable logic devices capable of being initialized?
    Are all programmable logic devices capable of being tristated?
    Can all on-chip oscillators be disabled?

LSI/VLSI ASIC CHECKLIST

Partitioning

Have all circuit functions been partitioned into reasonably small functional blocks (or clusters)?
    Have all tristate control lines been made directly or indirectly controllable by the ATE?

Controllability

Have all critical lines (RESET, HOLD, TRAP, WAIT, etc.) been made directly or indirectly controllable by the ATE?

Visibility

Have all bus and status indication lines been made accessible to the ATE?

Initialization

Are all logic functions able to be placed in a known state with a MASTER CLEAR line or with a maximum of 16 clock cycles?

Synchronization

Have all oscillators been made directly controllable by the ATE?

Have provisions been made to synchronize the ATE and the UUT?

Are all clocks of differing phases and frequencies derived from a single master (controllable) clock?

Are all memory elements clocked by a derivative of the master clock?

Self-Tests

Has ROM been allocated for self-test provisions?

Does on-board ROM contain self-test routines?

Device Standardization

Have standard LSI and VLSI devices which are testable by the ATE been used wherever possible?

Is the number of different part types a minimum?

Have parts been selected which are well characterized in terms of failure modes?

Structured Device Guidelines

Have structured design guidelines been followed for all custom LSI and VLSI devices?

Have all structured design rules been satisfied?

Does the design contain only synchronous logic?

ANALOG GUIDELINES CHECKLIST

Adjustments

Have adjustable components been removed wherever possible?
Are multiple, interactive adjustments prohibited for production items?

Relays

Have relays been replaced with solid-state switches wherever possible?

Feedback Loops

Have all critical feedback loops been designed to be opened during testing diagnostic operations?
Does the design avoid external feedback loops?

Signal Interfaces

Are all signal interfaces at levels easily monitored by test equipment?

High Voltages

Have all high-voltage sections been designed with dividers to drive test points with safety and accuracy?

Metering

Have metering provisions been allocated and designed?

Test Points

Have test points been selected such that they do not affect signal performance and allow for quick and reliable tester access, including ground test points?

Is each test point adequately buffered or isolated from the main signal path?

Functional Modularity

Have analog circuits been designed in functional blocks?

Are analog circuits partitioned by frequency to ease tester compatibility?

Are circuits functionally complete without bias networks on some other assembly?

High-Frequency Circuits

Have all high-frequency analog circuits been designed to be ATE compatible?

ATE Compatibility

Are stimulus frequencies compatible with tester capabilities?

Are stimulus rise time or pulse width requirements compatible with tester capabilities?

Do response measurements involve frequencies compatible with tester capabilities?

Are response rise time or pulse width measurements compatible with tester capabilities?

Are stimulus amplitude requirements within the capability of the test equipment?

Microwave Equipment

Have all microwave circuits been designed for ATE compatibility?

Hybrid Circuits

Has testability been considered for all hybrid circuits?

MECHANICAL GUIDELINES CHECKLIST

Human Engineering

Is the test point at the rear of the cabinet and the readout on the front of the cabinet, thus requiring two technicians?

Packaging for Accessibility

Have the minimal number of screws been used to gain access for inspection and testing?

Quarter-Turn Fasteners

If EMI allows, quarter-turn fasteners are preferable.

Fuses in Accessible Location

Fuses should always be in easily accessible locations.

Drawer Slides

Can the unit be mounted on drawer slides?

Drawer/Assembly Accessibility

Are all components in drawers and assemblies accessible for testing?

Cables/Service Loops

Are service loops of proper (not excessive) lengths?
    Can cables be easily replaced?

Cable Origin

Are cables easy to trace in the assembly?

Operational Test Points

Are operational test points edge-connector mounted?

Visual Indicators

Are all visual indicators human engineered?

RFI Shielding

Does the equipment specification require RFI shielding?

Test Points/Connectors

Are existing I/O connectors used for the majority of test interfaces?

Are the number of input and output (I/O) pins in edge connectors or cable connectors compatible with the I/O capabilities of the test equipment?

Are connector pins arranged so that shorting of physically adjacent pins will cause minimum damage?

Are power and ground included in the I/O or test connector?

Ease of Disconnection

Are quick-disconnect-type connectors used?

Color-Coded Tabs for Critical PCBs

Are high-voltage PCBs flagged by colored tabs?

Reference to Mating Connectors

Is mating connector information marked next to all receptacles?

PCB Extenders

Do card extenders make all components accessible, as well as the wire side of the PCB?

Use of Empty Card Slots for Testing

Can empty card slots be used for operational test points?

Solderless Connections

Do all (designated) replaceable components use solderless connections?

Keyed Connector Receptacles

Are all connectors keyed to prevent disasters?
    Is defeatable keying used on each board in order to reduce the number of unique interface adapters required?

Standard PCB Layout

Is a standard grid layout used on boards to facilitate identification of components?
    Are all components oriented in the same direction (e.g., pin 1 always in the same position)?

Extension Pads

Are all off-grid signals and internal nodes in multilayer boards brought to test points on the surface of the board?

Use of Standard Fixtures and Test Equipment

Can existing fixtures and test equipment be used?

Registration Holes

Are registration holes provided for aligning the UUT to the test fixture?

Ground Points

Is it easy to connect and disconnect ground leads to the designed item?

Individual Leads

Can individual test points be routed to a single connector?

Modular Functions/Feedback Loops

Can a complete function be packaged on a single PCB, including pull-up resistors?

Solder Mask

Are masking techniques used to isolate functional test points?

Components on One Side

Are components placed on only one side of the PCB?

Spacing between Components

Is enough spacing provided between components to allow for clips and test probes?

Minimum Nominal Selection

If nominals have not been eliminated, is their range and quantity kept to a minimum?

Switch Guards

Sensitive controls should be guarded so they cannot be accidentally disturbed.

Conformal Coating

Has the effect of conformal coating on test and repair requirements been taken into consideration?

Critical State of Art Measurements

Have all critical measurements been flagged to manufacturing for costing purposes?

BUILT-IN TEST GUIDELINES CHECKLIST

BIT Circuit Allocation

Is BIT circuitry optimally allocated in hardware, software, and firmware?

BIT Circuitry Overhead

Does the BIT circuitry make maximum use of normal functional circuitry in order to minimize BIT circuitry overhead?

Is the additional weight attributed to BIT within stated constraints?

Is the additional volume attributed to BIT within stated constraints?

Is the failure rate contribution of BIT within stated constraints?

Is the additional power consumption attributable to BIT circuity within stated constraints?

Is the additional parts count due to BIT within stated constraints?

Building-Block Approach

Does the BIT circuitry use a building-block approach (e.g., all inputs to a function are verified before the function itself is tested)?

Tester Control of BIT

Can BIT circuitry in each item be exercised under control of the ATE?

SOFTWARE GUIDELINES CHECKLIST

Halting

Can the software be set for halt on error by the test operator?

Continuing

Can the software be set for completion despite the occurrence of errors?

Looping

Can the software be selected to loop on a test or subtest?

Isolating Faults

Does the software aid in the isolation of faults?

Collecting Historical Data

Can the software be selected to record historical data for failure analysis?

Does BIT software include a method of saving on-line test data for the analysis of intermittent failures and operational failures which are nonrepeatable in a maintenance environment?

Test Initiation

Can the test be initiated and terminated by control of the operator?

Test Repetition

Can the test be selected for a specific number of repetitions?

Test Sequence

Can the test sequence be defined and initiated by the test operator?

Test Evaluation

Have test evaluation tools been included, including operator assistance displays?

Hard Copy Output

Can the test result be output in a hard copy?

LRU Testing

Do test diagnostics provide information to the lowest replaceable unit?

Single Entry, Single Exit

Do subroutines contain single entry and exit points?

Partitioning

Has software been partitioned and have structured programming techniques been employed?

DOCUMENTATION GUIDELINES CHECKLIST

User's Manual

Is it written?
    Are contents complete?

Maintenance Manual

Is it written?
    Are contents complete?

Schematic Diagrams

Are they present where required?

Relevant Waveforms, Timing Diagrams, and Logic Diagrams

Are they present where required?
    Have they been transferred to test engineering?

Wiring Diagrams and Wiring Run Lists

Are they present where required?

Assembly Drawings and Parts List

Are they available as required?

UUT Functional Description and Theory of Operation

Have they been written?
    Have they been transferred to test engineering?

Voltage/Resistance Chart for UUT Nodes

Are they available for analog circuits?

List of Test Equipment Required

Is it available?

UUT Performance Specifications and Test Procedures

Nothing is done until the paperwork is done!!

Appendix B:

Digital T-Score Rating System

To evaluate the testability of your digital PCB, perform the following steps. Use the score sheets contained within the procedures or make facsimiles of them to keep track of the score for each PCB to be evaluated. The last step contains a table which is used to convert your final score into a testability rating. This rating is a figure of merit which reflects the relative ease of testing and diagnosing faults on the PCB.

PERCENT NODES ACCESSIBLE

Accessible nodes are defined as nodes that are either available directly at a connector or through a maximum of one level of intervening combinatorial logic.

1. On the schematic diagram of the PCB under evaluation, trace an input lead to all of its termination points within the circuit. Mark the circuit path as you trace it so as to avoid retracing it later (Figure B-1). Exclude power and ground busses.

2. Now count the number of circuit packages tied to that input lead (see Figure B-1).

3. On the node accessibility score sheet (see Figure B-2), using the upper half of the form (labeled "access"), place a mark in the box whose numbered heading agrees with the number of circuit packages tied to the input lead you traced.

Note: In a later step you will count the number of marks you made in each box. Make your marks so that they can be readily grouped with others in multiples of 5 or 10 for easy counting.

4. Repeat steps 1 through 3 for each remaining input lead on the schematic diagram.

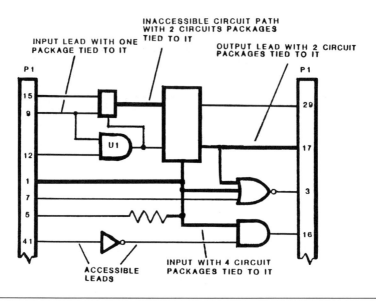

FIGURE B-1.   Node accessibility example.

5. Now repeat steps 1 through 3 for each output lead on the schematic diagram (see Figure B-1).

Note: At this point you should have every input and every output lead traced and marked off on the schematic. These leads and their corresponding nodes are termed *accessible,* since all of them can be accessed for testing purposes from an input or output pin on the PCB. Hence, the label *access* on the upper half of the score sheet. The remaining leads and circuit paths and their corresponding nodes which you have not yet marked off on the schematic are termed *inaccessible* and will be counted on the lower half of the score sheet.

6. On the schematic trace one of the inaccessible circuit paths to each of its termination points within the circuit. Mark the circuit path as you trace it so as to avoid retracing it later.

7. Now count the number of circuit packages tied to that circuit path (see Figure B-1).

8. On the node accessibility score sheet (see Figure B-2), using the lower half of the form (labeled "no access"), place a mark in the box whose numbered heading agrees with the number of circuit packages tied to the circuit path you traced.

9. Now repeat steps 6 through 8 for each remaining inaccessible circuit path on the schematic diagram.

10. Total the number of accessible nodes (connected circuit packages) and record their number under "total nodes (accessible)" on the upper half of the score sheet.

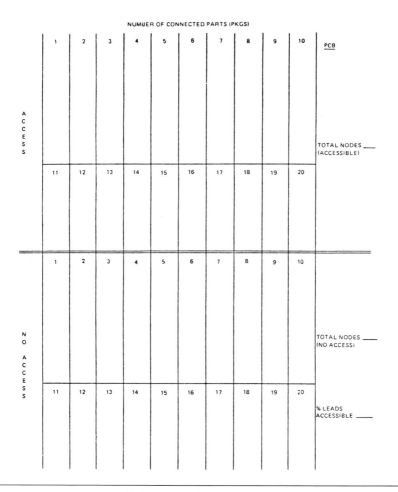

FIGURE B-2.   Node accessibility score sheet.

Note: For example, if you had five marks in the "6" number of connected parts box and two marks in the "4" number of connected parts box, you would have a total node count of (5 × 6) + (2 × 4) = 38.

11. Total the number of inaccessible nodes. Record this total under "total nodes (no access)" on the lower half of the score sheet.

12. Calculate the percentage of nodes accessible using the following formula:

$$\% \text{ Nodes accessible} = \frac{\text{total nodes (accessible)}}{\text{total nodes}} \times 100$$

13. Record the calculated percentage on the node accessibility score sheet under "% leads accessible."

14. Convert this calculated percentage to a weighted value using the following list.

% Leads Accessible	Weighted Value
91 to 100	30
81 to 100	27
71 to 100	24
61 to 100	21
51 to 100	18
41 to 100	15
31 to 100	12
21 to 100	8
11 to 100	4
0 to 100	0

15. Enter the weighted value on the PCB testability evaluation score sheet (see Figure B-3) opposite factor B1 in the "actual rating" column.

PROPER DOCUMENTATION

16. Now, examine the documentation supporting the PCB under evaluation. Does it meet or exceed the requirements listed in the following table? For each item, assign the given number of points when requirements are met or exceeded. Note the points assigned on a separate sheet of paper for later totaling.

Documentation Requirements	Rating
Logic diagrams or schematics of all detailed subcircuits (such as vendor-supplied modules mounted on the PCB) are provided either on the actual schematic diagram or as individual parts specification sheets	4
Detailed performance spec (with signal I/O tolerances) is provided (equivalent to MIL-STD-1519, dated August 1977).	8
Functional characteristics of each digital IC circuit type is available from vendor-supplied catalogs or on detailed drawings provided.	3
Functional designations are shown next to each pin number of all digital ICs on the schematic.	5
Power supply circuits on non-power-supply PCBs are shown in a single location on the schematic, and all voltages are labeled.	3

FAC-TOR	DESCRIPTION	SCORE	ACTUAL RATING	COMMENTS
B1	Percent Nodes Accessible			
B2	Proper Documentation			
B3	% of Sequential Circuits			
B4	PCB Complexity Count			
	Total Basic Score			
N1	Monostable Circuits			
N2	Counter (packages x stages)			
N3	Max. # of Function Blocks per Node (No Access)			
N4	Max. # of Function Blocks per Node (Accessible)			
N5	Seq. Supply Voltages			
N6	Non-Remov. Memories			
N7	Non-Rem. Buried Memory			
N8	Removable Complex Parts			
N9	Non-Rem. U-Proc, VLSI			
N10	Init. of Seq. CKTS			
N11	Ext. Loading Req'd			
N12	Different Logic Types			
N13	Buried Seq. Logic			
N15	Excess Warm-up Time			
N16	Tolerance			
N17	High Power			
N18	Critical Frequency			
N19	Clock Lines			
N20	Ext. Test Equipment			
N21	Environmental			
N22	Adjustments			
N23	Complex Signals			
N24	Redundant Logic			
N25	# of Logic Voltages			
N26	# of Power Supplies			
N27	Schematic Connectivities			
N28	I/O pin - Schematic			
N29	Dual Pin Designations			
N30	Symbols on Schematic			
	Total Negative Score			
	Net Total Score			

FIGURE B-3.   PCB testability evaluation score sheet.

Schematic references assembly diagram and gives the   2
part number of the next higher assembly.

17. Total all points awarded in step 16 and enter this figure on the PCB testability evaluation score sheet opposite factor B2 in the "actual rating" column.

PERCENT OF SEQUENTIAL CIRCUITS

Note: In the following step, you will need to know the difference between a sequential IC circuit and a combinatorial IC circuit. *Combinatorial ICs* are those such as gates, multiplexers, and the like, where the outputs of the circuits depend only on the present inputs. *Sequential ICs* are those such as flip-flops, counters, shift registers, memories (where clocking is involved), and so on, where the output depends not only on the present inputs but also on past inputs or sequences of inputs.

18. Using the PCB parts list, total the number of sequential IC packages.

19. Divide the number of sequential IC packages by the total of IC packages contained on the PCB and multiply by 100. This is the percentage of sequential IC circuits. Enter this figure opposite factor B3 in the "score" column of the score sheet.

20. Convert this percentage into an actual rating using the following list.

Sequential ICs (%)	Actual Rating
0 to 14	15
15 to 24	10
25 to 39	5
40 to 49	3
50 to 100	0

21. Enter the actual rating on the score sheet opposite factor B3 in the actual rating column.

PCB COMPLEXITY COUNT

22. Use the following list to determine the complexity count for all sequential IC circuit parts shown on the schematic. Ignore combinatorial ICs. For each sequential IC, find the point value in the table and record it on a separate sheet for later totaling.

Sequential IC Type	Point Value
Flip-flop	7
Latch	7
4-bit shift register	35
4-bit counter	35
Hex D register	50
8-bit shift register	70

8-bit counter	70
Memory (RAM) <16K	100
Memory (RAM) 16K and above	500
μP, LSI, VLSI (with BIT)	500
μP, LSI, VLSI (w/o BIT)	1000

Note: The point value for most other sequential ICs can be extrapolated from this list.

23. Total the complexity points noted for each IC part. Enter this total complexity point value opposite factor B4 in the score column.

24. Convert the total complexity point value to an actual rating.

Total Complexity Point Value	*Actual Rating*
Less than 300	30
301 to 500	24
501 to 800	18
801 to 500	12
1201 to 1200	6
1801 or higher	0

25. Enter the actual rating opposite factor B4.

26. Add all of the percentage points in the actual rating column for factors B1 through B4. This is your total basic score for all positive (easy to test) factors. Enter this figure just below the factor B4 score in the actual rating column.

You have now completed compiling your score for all positive factors. The rest of the factors you will be working with involve negative (difficult to test) aspects of your PCB. These will later be subtracted from your total basic score score to obtain the net score.

MONOSTABLE CIRCUITS

27. Does your PCB have any monostable circuits? If so, continue to step 28. If not, enter 0 as your actual rating opposite factor N1 on the score sheet and go to step 30.

28. Examine each monostable circuit on the PCB and assign point values according to the list below. Note each point value on a separate sheet for later totaling. No points are assigned for characteristics not listed here.

Scoring Factors	*Point Value (each)*
If the input is not readily controllable	5
If the output is not accessible	5

If the output cannot be replaced by the   10
  ATE
If parametric measurements are required   1

29. Total the points noted for each monostable circuit. Enter this value opposite factor N1 in the actual rating column of the score sheet.

COUNTERS

30. Does your PCB have any counter circuits? If so, continue to step 31. If not, enter 0 as your actual rating opposite factor N2 on the score sheet and go to step 35.

NOTE: In the following four steps, you will be examining each counter circuit in your schematic diagram. For these steps, a *counter circuit* is defined as a series of individual counter stages between non-counter circuits. In the example (see Figure B-4), we have a counter circuit of three stages. In turn, the three stages comprise two individual IC packages, U1 and U2 (U1A and U1B are part of the same package). The example counter circuit is an accessible circuit since it can be controlled and monitored via I/O pins through no more than one level of gating (e.g., AND, OR, NOT functions).

Definition: 1 stage = 4-bit counter.

31. Examine any counter circuit on the schematic. Determine the number of stages, number of individual IC packages, and whether the circuit is accessible or inaccessible. Multiply the number of counter

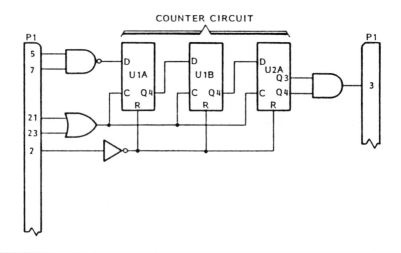

FIGURE B-4.   Counter stage example.

stages by the number of individual IC packages (3 stages × 2 IC packages in the Figure B-4 example) to determine the point value for your counter circuit (6 in this example).

32. Convert this point value to an actual rating using the following list (N is the counter point value). Note the actual rating on a separate sheet for later totaling.

Note: As a general design guideline, provide ATE breakpoints after no more than every four counter stages.

Counter Point Value	Actual Rating
Less than 5	0
5 to 9 accessible	2
5 to 9 inaccessible	3
10 or more accesible	$4 + [0.05 \times (N - 10)^2]$
10 or more inaccessible	$5 + [0.1 \times (N - 10)^2]$

Note: For example, if your counter point value was 12 and your circuit was inaccessible, your actual rating would be $5 + [0.1 \times (12 - 10)^2] = 5.4$.

33. Repeat steps 31 and 32 for each remaining counter circuit on your schematic.

34. Total the actual rating noted for all counter circuits on your schematic. Enter this total opposite factor N2 in the actual rating column.

MAXIMUM NUMBER OF FUNCTION BLOCKS PER NODE (NO ACCESS)

35. Refer to the node accessibility score sheet you filled in previously (see Figure B-2). Look for the number of marks made in each "no access" box.

36. Assign actual rating points to each no access box as follows: Multiply the number of marks in each no access box by the point value for that box obtained from the following list. Note this actual rating point value for each box on a separate sheet for later totaling.

No Access Box Number	Point Value
1–3	0
4	0.1
5	0.2
6	0.5
7	1.0

8	1.3
9	1.7
10–20	2.0

Note: For example, if you had seven marks in the "6" no access box, you would have an actual rating point value of 7 × 0.5, or 3.5, for that box.

37. Total the actual rating points noted for all no access boxes. Enter this total actual rating point value opposite factor N3 in the actual rating column.

MAXIMUM NUMBER OF FUNCTION BLOCKS PER NODE (ACCESS)

38. Again, refer back to the node accessibility score sheet filled out previously. This time, look for the number of marks made in each "access" box.

39. Assign actual rating points to each access box as follows: Multiply the number of marks in each access box by the point value for that box obtained from the following list. Note this actual rating point value for each box on a separate sheet for later totaling.

No Access Box Number	Point Value
1–4	0
5	0.1
6	0.2
7	0.5
8	0.6
9	0.8
10–20	1.0

40. Total the actual rating points noted for all access boxes. Enter this total opposite factor N4 in the actual rating column.

SEQUENCE SUPPLY VOLTAGES

41. Does your PCB require two or more supply voltages which need a special turn-on or turn-off sequence? If so, enter 10 as your actual rating opposite factor N5. If not enter 0.

NONREMOVABLE MEMORIES

42. Does the PCB have any memory ICs which are permanently wired or soldered to the PCB, as opposed to plugged into a socket, or that cannot be electrically tristated via an edge connector pin? If so, continue to step 43. If not, enter 0 opposite factors N6 and N7 in the actual rating column and go to step 47.

Note: In the following four steps, an *accessible* memory IC is defined as one with all of its I/O leads accessible to connector pins within one level of logic. A memory IC is considered *inaccessible* if it has one or more of its I/O leads inaccessible to PCB pins.

43. Examine the memory size of each accessible memory IC permanently wired or soldered to the PCB. Ignore plug-in memories. Assign points to each IC according to the following list. Note the point value for each memory IC on a separate sheet for later totaling. If you have no accessible memory ICs, enter 0 opposite factor N6 in the actual rating and go to step 45.

Memory Size (bits)	Point Value
100K or greater	10
32K to 99K	6
8K to 31K	4
1K to 7K	2

44. Total the point values for all accessible memory ICs examined. Enter this total opposite factor N6 in the actual rating column.

NONREMOVABLE BURIED MEMORIES

45. Examine the memory size of each inaccessible memory IC permanently wired or soldered to the PCB. Ignore plug-in memories. Assign points to each IC according to the following list. Note the point value for each memory IC on a separate sheet for later totaling. If you have no inaccessible memory ICs, enter 0 opposite factor N7 in the actual rating column and go to step 47.

Memory Size (bits)	Point Value
Less than 1K	10
1K or greater	20

Note: If address and data lines are available through test points or through BIT, which allows for fault isolation to a failed component, proceed to step 47 and enter 0 opposite factor N7.

46. Total the point values for all inaccessible memory ICs examined. Enter this total value opposite factor N7 in the actual rating column of the score sheet.

REMOVABLE COMPLEX PARTS

47. Does your PCB have any socket-mounted or plug-in VLSI ICs, microprocessors, or memories that must be removed prior to automatic digital testing? If so, go to step 48. If not, enter 0 opposite factor N8 in the actual rating column and go to step 49.

48. For each such component that must be removed prior to testing, assign 1 point. Total the points and enter this value opposite factor N8 in the actual rating column.

NONREMOVABLE COMPLEX PARTS

49. Does your PCB have any VLSI ICs, microprocessors, or memories which are permanently wired or soldered to the PCB, as opposed to plugged into a socket? If so, go to step 50. If not, enter 0 opposite factor N9 in the actual rating column and go to step 52.

Note: In the following two steps, an *accessible* component is defined as one with all of its I/O leads accessible to PCB pins. A component is considered *inaccessible* if it has one or more of its I/O leads inaccessible to PCB pins.

50. Examine each complex component permanently wired or soldered to the PCB for accessibility. Ignore plug-in components. Assign points to each complex component according to the following list. Note the point value for each component on a separate sheet for later totaling. If a faulty component on a PCB using VLSI/microprocessor chips can be fault isolated by the use of on-chip or on-board BIT, or external ATE, put 0 on factor N9 in the actual rating column.

Component Accessibility	Point Value
Accessible	10
Inaccessible	100

51. Total the points for all microprocessors, VLSI ICs, and other complex components examined. Enter this total opposite factor N9 in the actual rating column.

INITIALIZATION OF SEQUENTIAL CIRCUITS

52. Examine all sequential ICs (such as counters, flip-flops, memories, etc.) on the PCB that have reset or set capabilities. For those which do, are they all resettable or presettable via a PCB connector pin by applying a digital stimulus pattern less than 16 bits in length to that pin? If so, enter 0 as your actual rating opposite factor N10 and go to step 55. If not, continue to step 53.

53. Assign points to each nonresettable and nonpresettable (via PCB connector pin) sequential IC (assuming the IC is so capable) according to the list. Note each point value on a separate sheet for later totaling.

Reset/Set Factors	*Point Value (each)*
Can be either set or reset via PCB connector pin via a pattern no more than 16 bits in length	0
Can be neither set nor reset via PCB connector pin via a pattern no more than 16 bits in length	5

54. Total the point values noted for all nonresettable or nonpresettable (via PCB connector pin) sequential ICs. Enter this value opposite factor N10 in the actual rating column.

EXTERNAL LOADING REQUIRED

55. Does your PCB require external loads (such as pull-up resistors, resistive loads, or transistorized loads) to be added to the automatic tester? If so, continue to step 56. If not, enter 0 as your actual rating opposite factor N11 and go to step 58.

56. Determine the actual rating for your PCB using this list:

Counter Point Value	*Actual Rating*
1–49 passive components such as resistors	2
50 or more paasive components	3
5 or more active components, such as transistor/ resistor combinations	5

57. Enter the actual rating opposite factor N11 in the actual rating column.

DIFFERENT LOGIC TYPES

58. Look at the parts list for the PCB and determine the number of different types of ICs used on the PCB. Assign an actual rating to your PCB based on the number of different IC types, using the following list.

Number of IC Types	Actual Rating
<10	0
10	1
>10	1 + (1 for each additional 3 types above 10)

59. Enter the actual rating opposite factor N12 in the actual rating column.

BURIED SEQUENTIAL LOGIC

60. Examine each cluster of inaccessible sequential ICs other than counter circuits and memories with and without one level of buffer circuits on the PCB schematic diagram. Count the ICs in each cluster and assign point values to each cluster according to the following list. Note the point value for each cluster on a separate sheet for later totaling.

Number of ICs in Cluster	Point Value
1 or 2	0
3 or 4	1
5 to 10	5
> 10	10

Note: A cluster is (three or more) components connected to perform a sequential function.

61. Total the points noted for all inaccessible sequential IC clusters on the PCB. Enter this total opposite factor N13 in the actual rating column.

EXCESS WARM-UP TIME

62. Does the time for your PCB to warm up and stabilize exceed 3 minutes? If so, enter a 3 as your actual rating opposite factor N15 and go to step 35. If not, enter 0.

TEST EQUIPMENT/PCB TOLERANCE

63. Check the measurement capability of the recommended test equipment to be used to test your PCB. If the measurement capability of all pieces of test equipment is at least 10 times more accurate than that required by your PCB, enter 0 as your actual rating opposite factor N16 on the score sheet and go to step 66. If any piece of the test equipment's measurement capability is not at least 10 times more accurate than required by the PCB, go to step 64.

64. For test equipment measurement deficiencies, assign an actual rating to your PCB according to the following list.

Total Complexity Point Value	Actual Rating
3 to 9 times more accurate than required by the PCB	2
Less than 3 times more accurate than required by the PCB	5

65. Enter the actual rating opposite factor N16 in the actual rating column.

HIGH POWER

66. Does the PCB require any high voltages or currents in order to be tested? If so, continue to step 67. If not, enter 0 as your actual rating opposite factor N17 and go to step 69.

67. For each high voltage or current required by the PCB, assign point values according to the list. Note each point value on a separate sheet for later totaling.

Voltage or Current Required	Point Value
More than 5 A of current	5
High voltage greater than 110 V p-p	2
Multiple parallel pins for high current	1

68. Total the point values noted for all high voltages or currents. Enter this total opposite factor N17 in the actual rating column.

CRITICAL FREQUENCY

69. Does the PCB have any critical frequencies that must be measured or supplied, or does it require coaxial cable in the automatic tester? If so, continue to step 70. If not, enter 0 as your actual rating opposite factor N18 and go to step 72.

70. For critical frequencies or coax requirements, assign an actual rating to the PCB according to the list.

Critical Frequency or Coax Requirement	Actual Rating
Requires coax in ATE	5
Critical frequency greater than 10 MHz	3
Critical frequency of 5–10 MHz	2
Critical frequency of 1–4 MHz	1

Note: Do not add ratings. Assign only the highest rating possible. For instance, if the PCB has a critical frequency of 15 MHz and requires coax in the tester, assign a rating of 5 to the PCB.

71. Enter the actual rating opposite factor N18 in the actual rating column.

CLOCK LINES

72. Does the PCB have any clock oscillators on it? If so, continue on to step 73. If not, enter 0 as your actual rating opposite factor N19 and go to step 75.

Note: A clock oscillator is most easily tested if it can be disabled by an external signal and an external clock substituted for the PCB clock. The clock oscillator output should also be accessible for monitoring. Points (penalties) are therefore assigned based on these factors. In the following step and table, externally controlled refers to the ability of the on-board oscillator to be externally disabled and the board circuitry capable of being driven by an external clock.

73. Assign an actual rating to the PCB based on the number and accessibility of your PCB clock oscillators according to the following list.

Factor	Actual Rating
Each clock oscillator externally controlled with output accessible	0
Each clock oscillator externally controlled with output inaccessible	20

| Each clock oscillator not externally controllable with output accessible | 50 |
| Each clock oscillator not externally controllable with output inaccessible | 70 |

74. Enter the actual rating opposite factor N19 in the actual rating column.

EXTERNAL TEST EQUIPMENT

75. In order to be fully tested, does the PCB require additional pieces of test equipment other than the automatic tester? If so, continue to step 76. If not, enter 0 as your actual rating opposite factor N20 on the score sheet and go to step 78.

76. For additional test equipment required, assign an actual rating according to the following list. The number of pieces of test equipment is N.

Pieces of Additional Equipment   *Actual Rating*

$$N \longrightarrow 2(N \times N)$$

77. Enter the actual rating opposite factor N20 in the actual rating column.

ENVIRONMENTAL

78. Are special environmental chambers required to test the PCB? If so, continue to step 79. If not, enter 0 as your actual rating opposite factor N21 and go to step 81.

79. Assign an actual rating to the PCB based on type of environmental chamber or area required according to the following list.

Conditions Required	*Actual Rating*
Forced air (i.e., fan-cooled) or cold or hot chamber area	2
Altitude-controlled, or electromagnetic interference (EMI) isolated chamber	10

80. Enter the actual rating opposite factor N21 in the actual rating column.

ADJUSTMENTS

81. Are there any adjustments that are required to be made to the PCB during the test period? If so, continue to step 82. If not, enter 0 as your actual rating opposite factor N22 and go to step 84.

82. For each adjustment to be made to the PCB during testing, assign 2 points. For each interactive adjustment required (i.e., two or more pots or variable capacitors that must be adjusted and readjusted until no further adjustments between them are necessary), assign 4 points. Note each point value on a separate sheet for later totaling.

83. Total the points assigned and enter opposite factor N22 in the actual rating column.

COMPLEX SIGNALS

84. Does the PCB have any complex or unusual wave shapes, either as inputs or outputs, which require interpretation by the test operator? Interpretation may be of such items as phase, frequency, rise time, pulse width, and so on. If so, continue to step 85. If not, enter 0 as your actual rating opposite factor N23 and go to step 87.

85. For each complex or unusual wave shape that requires interpretation, assign 2 points. For each instance where two or more coincident complex or unusual wave shapes occur, assign 5 points. Note each point value on a separate sheet for later totaling.

86. Total the points above and enter this opposite factor N23 in the actual rating column.

REDUNDANT LOGIC

87. Does the PCB have any groups of two or more parallel redundant logic signals which, because of their being parallel and not in the same package, prevent fault isolation to individual logic circuit failures?

Note: Do not count those parallel logic circuits which can be individually fault isolated by PCB BITE circuits.

88. If the PCB does have parallel redundant logic circuits as described above, go to step 89. If not, enter 0 as your actual rating opposite factor N24 and go to step 90.

89. For each non-fault-isolatable group of parallel logic circuits, assign points according to the following list. Note each point value on a separate sheet for later totaling.

Number of Parallel Circuits	*Point Value*
2	2
3 or more	3

90. Total all points and enter opposite factor N24 in the actual rating column.

NUMBER OF LOGIC FAMILIES

91. Does the PCB have more than one type of logic family? If so, go to step 92. If not, enter 0 as your actual rating opposite factor N25 and go to step 93.

92. For each type of logic family over one, assign 5 points per logic family. Total all points and enter opposite factor N25 in the actual rating column.

NUMBER OF POWER SUPPLIES

93. Determine the number of separate power sources which must be supplied by the test station to power the PCB. Power supplies may be internal to the automatic digital tester or supplied as separate pieces of test equipment.

94. If the number of separate power supplies is three or less, enter 0 as your actual rating opposite factor N26 and go to step 96. If not, go to step 95.

95. If the number of separate power supplies is four or more, assign 1 point for each additional power supply beyond three. Enter the total opposite factor N26 in the actual rating column.

Note: For example, if the number of separate power sources required to be supplied to the PCB were 8, your actual rating would be 8 − 3, or 5.

SCHEMATIC CONNECTIVES

96. Examine the schematic for the PCB. Check for the following: Is the schematic contained on a single page? If it is spread over many pages, do all interpage signals show sheet numbers(s) and zone(s) where the signals go to or come from (see Figure B-5)?

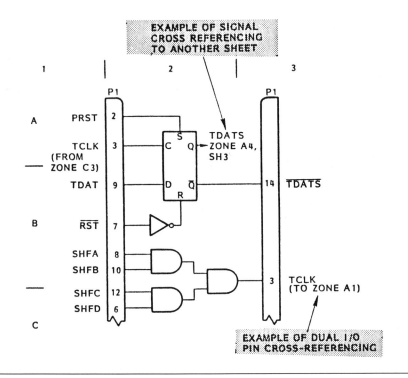

FIGURE B-5.   Schematic example.

97. If the schematic is not as described above, enter 20 as your actual rating opposite factor N27. If the schematic conforms to the above description, enter 0.

I/O PIN SCHEMATIC

98. Examine the schematic for the PCB. Are all input pins shown on the left side of the diagram and output pins on right, with no I/O pins shown in the middle? If not, enter 2 points per signal that does not conform.

99. If any I/O pins are shown in the middle of the schematic, enter 3 points/signal as your actual rating opposite factor N28. If the schematic conforms to the step 98 description, enter 0.

DUAL PIN DESIGNATIONS

100. Examine the schematic for the PCB. Check for I/O pins which are shown more than once on the same sheet without being cross-referenced (see Figure 9-7).

101. If there are no I/O pins shown more than once on the same sheet without being cross-referenced, enter 0 as your actual rating opposite factor N29 and go to step 104. If duplicate I/O pins are not cross-referenced, go to step 102.

102. For each instance of duplicate I/O pins not cross-referenced, assign 3 points. Total all points assigned and enter opposite factor N29 in the actual rating column.

SYMBOLS ON SCHEMATIC

103. Examine the schematic for the PCB. Only a single symbol should be used to depict a specific component or hardware part. This includes ICs, transistors, resistors, capacitors, I/O connectors, LEDs, and so on. However, functional variations for logic components may be used.

104. You have now finished checking the PCB and its documentation for all negative (hard-to-test) factors. Total all percentage points in the actual rating column for factors N1through N30.

105. Enter this total negative score on the appropriately labeled line in the actual rating column of the score sheet.

106. Subtract this total negative score from the total basic score you compiled in step 26. Enter this result as your net total score on the score sheet.

107. Now use your net total score to determine the testability of the PCB according to the list below. Naturally, the higher the score, the easier the PCB will be to test. However, if your score falls below 31, the PCB or its documentation should be redesigned or reworked to raise the score.

	Net Total Score (%)	*PCB Testability*
	81 to 100	Very easy
Acceptable	66 to 80	Easy
	46 to 65	Medium easy
	31 to 45	Average
	11 to 30	Hard
Redesign	1 to 10	Very hard
	0 or less	Impossible to test and troubleshoot without cost penalties

Bibliography

Bardell, P. H., and W. H. McAnney, "Self-testing of multichip logic modules," Proc. IEEE International Test Conference, 1982, pp. 200–204.

Batson, J., *In-Circuit Testing*, Van Nostrand Reinhold, New York, 1985.

Bennetts, R. G., *Design of Testable Logic Circuits*, Addison-Wesley, 1984.

Bottorff, P. S., S. DasGupta, R. G. Walther, and T. W. Williams, "Self-testing using shift-register latches," *IBM Technical Disclosure Bulletin*, Vol. 25, No. 10, March 1983, pp. 4958–4960.

Consolla, W., and F. Danner, "An Objective Printed Circuit Board Testability Design Guide and Rating System," RADC Report Number RADC-TR-79-327.

DasGupta, S., et al., "Chip partitioning aid: a design technique for partitionability and testability in VLSI," Proc. ACM/IEEE Design Automation Conference, 1978, pp. 203–208.

DasGupta, S., et al., "An enhancement to LSSD and some applications of LSSD in reliability, availability and serviceability," Proc. IEEE Fault Tolerant Computing Symposium, 1980, pp. 32–34.

DeSena, A., "Testability is now possible without additional real estate," *Computer Design*, November 15, 1987.

Eichelberger, E. B., and T. W. Williams, "A logic design structure for LSI testability," *Journal of Design Automation and Fault-Tolerant Computing*, Vol. 2, No. 2, May 1978, pp. 165–178.

"The primer of high-performance in-circuit testing," Factron Schlumberger, Wimborne, Dorset, U.K., 1985.

Frohwerk, R. A., "Signature analysis: a new digital field service method, *Hewlett Packard Journal*, Vol. 28, No. 9, May 1977, pp. 2–8.

Fujiwara, H., *Logic Testing and Design for Testability*, MIT Press, Cambridge, MA, 1985.

Goel, P., "Test costs analysis and projections," IEEE Design Automation Conference, 1980, pp. 77–82.

Goering, R., "Boundary-scan technique targets board-level testability," *Computer Design*, October 1, 1987, pp. 47, 49.

IEEE P1149 Standard Testability Bus Specification, Draft 8.

IEEE P1149.1 Standard Test Access Port and Boundary-Scan Architecture, Draft 6, November 22, 1989.

IEEE P1149.2 Extended Serial Digital Testability Bus Protocol, Draft.

IEEE P1149.3 Real Time Digital Testability Bus Protocol, Draft.

IEEE P1149.4 Real Time Analog Testability Bus Protocol, Draft.

Joselyn, L., "ATE vendors slow to accept DFT," *TEST Magazine*, June 1988.

Komonytsky, D., "LSI self-test using level-sensitive scan design and signature analysis," Proc. IEEE International Test Conference, 1982, pp. 414–424.

Konemann, B., et al., "Built-in logic block observation techniques," Proc. IEEE 1979 Test Conference, 1979, pp. 37–41.

Kuban, J.R., and J. E. Salick, "Testing approaches in the MC68020," *VLSI Design*, Vol. 5, No. 11, 1984, pp. 22–30.

Lake, R., "A fast 20K gate array with on-chip test system," *VLSI Systems Design*, June 1986, pp. 46–55.

Laurent, D., "An example of test strategy for computer implemented with VLSI circuits," Proc. IEEE International Conference on Circuits and Computers, 1985, pp. 679–682.

Leibson, S., "Design for testability creates better products at lower cost," *EDN Magazine*, March 31, 1988.

Markowitz, M., "High-density ICs need design-for-test methods," *EDN Magazine*, November 24, 1988.

McCluskey, E., *Logic Design Principles* (with emphasis on testable semicustom circuits), Prentice Hall, Englewood Cliffs, NJ, 1986.

Miller, M., S. Lewis, and P. Rangnekar, "On-chip serial diagnostics simplifies VLSI system design and testing," *Integrated Device Technology*, April 1987.

Parker, K., *Integrating Design and Test: Using CAE Tools for ATE Programming*, IEEE Computer Society Press, 1987.

Pradhan, M. M., et al., "Developing a standard for boundary-scan implementation," Proc. IEEE International Conference on Computer Design, 1987, pp. 462–466.

Resnick, D. R., and A. G. Bell, "Real world built-in test for VLSI," Proc. IEEE COMPCON, 1986, pp. 436–440.

Rosenberg, L., "The evolution of design automation toward VLSI," Proceedings of the 17th Design Automation Conference, IEEE.

Sideris, G., "The drive for IC test-bus standards," *Electronics*, 11 June 1987, pp. 68–71.

Strassberg, D., "Pioneering engineers begin to adopt board-level automatic test generation," *EDN Magazine*, March 17, 1988.

Tsui, F., *LSI/VLSI Testability Design*, McGraw-Hill, New York, 1987.

Tunic, D., "Design-to-test links foster IC development," *Electronic Design*, October 29, 1987.

Turino, J., *Design for Testability*, Logical Solutions, Inc., 1979.

Turino, J., "Forecast '84—trends in LSI/VLSI testing," *Semiconductor International*, January 1984.

Turino, J., "Design of SMA for testability requires new rules," *Electronic Packaging and Production*, September 1984.

Turino, J., "Functional testing's place in electronics manufacturing," *Evaluation Engineering*, September 1984.

Turino, J., "ATE costs are swelling as chips shrink," *Electronics Week*, October 8, 1984.

Turino, J., "Forecast '85—coming major changes in test," *Semiconductor International*, January 1985.

Turino, J., "Enhancing built-in test on SMT boards," *Evaluation Engineering*, June 1985.

Turino, J., "Testability circuit solves SMT board access problems," *Electronic Packaging and Production*, January, 1986.

Turino, J., "Test/evaluation outlook for '87," *Evaluation Engineering*, December 1986.

Turino, J., "A proposed standard testability bus," *Evaluation Engineering*, October 1987.

Turino, J., "Test/evaluation outlook for '88," *Evaluation Engineering*, December 1987.

Turino, J., "Design for testability—a competitive strategy," *Evaluation Engineering*, March 1988.

Turino, J., "Circuit testability is critical for product success," *EDN Magazine*, September 15, 1988.

Turino, J., "Design for test is a must," *New Electronics*, August 1988.

Turino, J., "ATE outlook For 1989," *Evaluation Engineering*, December 1988.

Turino, J., and H. Frank Binnendyk, *Design To Test*, Logical Solutions, Inc., 1982.

Turino, J., and David Mei, *Microporcessor Board Testability*, Logical Solutions, Inc., 1980.

United States Patent Number 4,720,672.

VHSIC Phase 2 Interoperability standards: ETM-Bus specification.

Weiss, D., "VLSI test methodology," *Hewlett-Packard Journal*, September 1987, pp. 24–25.

Wagner, P. T., "Interconnect testing with boundary scan," Proc. IEEE International Test Conference, 1987, pp. 52–57. (Includes simple methods of defining boundary scan tests for board interconnect and provides estimates of test length.)

Williams, T. W., and E. B. Eichelberger, "Random patterns within a structured sequential logic design," Proc. IEEE Semiconductor Test Symposium, 1977, pp. 19–27.

Williams, T. W., and K. Parker, "Design for testability—a survey," IEEE Transactions on Computers, January 1982.

Wittenberg, R., "Driving the testability bus," EE Times, October 12, 1987.

Zasio, J. J., "Shifting away from probes for wafer test," Proc. IEEE COMPCON, 1983, pp. 395–398.

Index